INTERNET THEORY, TECHNOLOGY AND APPLICATIONS

FUTURE OF THE INTERNET: SOCIAL NETWORKS, POLICY ISSUES AND LEARNING TOOLS

INTERNET THEORY, TECHNOLOGY AND APPLICATIONS

Additional books in this series can be found on Nova's website under the Series tab.

Additional E-books in this series can be found on Nova's website under the E-books tab.

FUTURE OF THE INTERNET: SOCIAL NETWORKS, POLICY ISSUES AND LEARNING TOOLS

RICK D. SULLIVAN

AND

DOMINICK P. BARTELL

EDITORS

Nova Science Publishers, Inc.

New York

Library of Congress Cataloging-in-Publication Data

Future of the Internet : social networks, policy issues, and learning tools
/ editors, Rick D. Sullivan and Dominick P. Bartell.
     p. cm.
 Includes bibliographical references and index.
 ISBN 978-1-61209-597-4 (hardcover : alk. paper)
 1. Internet. 2. Online social networks. 3. Telecommunication policy. 4.
Disclosure of information. I. Sullivan, Rick D. II. Bartell, Dominick P.
 TK5105.875.I57F8927 2011
 004.67'8--dc22
                                        2011001551

Published by Nova Science Publishers, Inc. † New York

CONTENTS

PREFACE

This new book examines the current aspects of the internet which affect society, as well as the projected effect the internet will have in the future. Topics discussed include using social networking sites to look at the future; personal information agency and reseller adherence to key privacy principles; laws regarding personal information, including social security numbers, being used by internet resellers; e-governance in the Pacific Islanders and the use of web 2.0 tools to support collaborative learning.

Chapter 1 - "Social networking" is not a new concept. As Barabási [1] puts it, various traditional facets of our existence might be reinterpreted as social networking, from early Christians to the World Wide Web. It refers to the act of relating nodes (e.g., individuals, organisations, other social entities) through social links (e.g., friendship, professional relationship, information exchange)[2]

Before the 20th century, social networking could only take place at short distance, or with considerable time delay when travelling or exchanging letters. Since the 1950s, telephones, tele- and videoconferencing have allowed real-time communications even over long distances. These two-way communication channels are generally good links for pairs of people or small groups, but they tend to be inefficient for large-scale, mass interactions. Moreover they enable only simultaneous communications.

Chapter 2 - In fiscal year 2005, the Departments of Justice, Homeland Security, and State and the Social Security Administration reported that they used personal information obtained from resellers for a variety of purposes. Components of the Department of Justice (the largest user of resellers) used such information in performing criminal investigations, locating witnesses and fugitives, researching assets held by individuals of interest, and detecting prescription drug fraud. The Department of Homeland Security used reseller information for immigration fraud detection and border screening programs. Uses by the Social Security Administration and the Department of State were to prevent and detect fraud, verify identity, and determine eligibility for benefits. The agencies spent approximately $30 million on contractual arrangements with resellers that enabled the acquisition and use of such information. About 91 percent of the planned fiscal year 2005 spending was for law enforcement (69 percent) or counterterrorism (22 percent).

Chapter 3 - Financial institutions such as banks, credit card companies, securities firms, and insurance companies use personal data obtained from information resellers to help make eligibility determinations, comply with legal requirements, prevent fraud, and market their products. For example, lenders rely on credit reports sold by the three nationwide credit bureaus to help decide whether to offer credit and on what terms. Some companies also use reseller products to comply with PATRIOT Act rules, to investigate fraud, and to identify customers with specific characteristics for marketing purposes.

Chapter 4 - We found 154 Internet information resellers with SSN-related services. Most of these resellers offered a range of personal information, such as dates of birth, drivers' license information, and telephone records. Many offered this information in packages, such as background checks and criminal checks. Most resellers also frequently identified individuals, businesses, attorneys, and financial institutions as their typical clients, and public or nonpublic sources, or both as their sources of information.

Chapter 5 - Many commentators have identified the Pacific Islands countries as being plagued by a diverse range of challenges - political insecurity, ethnic divisions, corruption, economic under-development and social inequality. Making these problems more difficult to tackle is the so-called 'tyranny of distance', the large physical distances and lack of communications infrastructure between the remote communities of the islands and their capitals.

Chapter 6 - The ability to collaborate is becoming more and more important in today's world in which tasks are getting more and more interdisciplinary and complicated to accomplish. It is therefore essential to prepare students on collaborative tasks while they are in schools so that they can become competent team workers when they enter the workforce. This paper illustrates how various web 2.0 tools have been used to support collaborative learning. The web 2.0 tools presented in this paper include Weblog, Wiki, Google Docs, Yahoo group, and Facebook. The affordances of these tools for collaborative learning together with examples of using the tools to support teachers and students in collaborative learning processes are also described.

In: Future of the Internet: Social Networks…   ISBN: 978-1-61209-597-4
Editors: Rick D. Sullivan and Dominick P. Bartell   ©2011 Nova Science Publishers, Inc.

Chapter 1

USING ONLINE SOCIAL NETWORKS TO LOOK AT THE FUTURE

Olivier Da Costa, Romina Cachia, Ramón Compañó and Effie Amanatidou*

AN OVERVIEW OF ONLINE SOCIAL NETWORKS (OSNS)

"Social networking" is not a new concept. As Barabási [1] puts it, various traditional facets of our existence might be reinterpreted as social networking, from early Christians to the World Wide Web. It refers to the act of relating nodes (e.g., individuals, organisations, other social entities) through social links (e.g., friendship, professional relationship, information exchange)[2]

Before the 20th century, social networking could only take place at short distance, or with considerable time delay when travelling or exchanging letters. Since the 1950s, telephones, tele- and videoconferencing have allowed real-time communications even over long distances. These two-way communication channels are generally good links for pairs of people or small groups, but they tend to be inefficient for large-scale, mass interactions. Moreover they enable only simultaneous communications.

Traditional mass communication takes place through (paper or online) newspapers, radios or television channels. However, these are one-way channels of information and communication. Although online newspapers do now include some multimedia ingredients (e.g. audio, graphics, animation, videos and interactive tools) to the traditional text, they still remain top-down information media not allowing two-way communication.

The recent emergence of the Internet and the development of appropriate software for social networking, have given new sense to large-scale communication. For the first time in media history, many-to-many communication has become possible. As Sack [3] argues, digital large-scale conversations have common characteristics, in that they are large, network-based and public. New modalities of interactions, currently referred to as "social computing",

---
* Corresponding author olivier.dacosta@m4x.org

are now enabled. As Wellman states, in recent years, computer systems have become "inherently social" connecting people and organisations [4], reflecting Castells' observation that we are shifting from group-based societies to networked societies [5]

Online Social Networks (OSNs) are one of the most remarkable ongoing technological and social phenomena, with several of them comprising millions of members from all over the world (e.g. MySpace [6], Facebook [7] Orkut [8] Friendster [9]) and being among the most visited websites globally [10] Some networks target large populations: teenagers and artists for MySpace, college students for Facebook. Others have a focus on business networking in general (e.g. LinkedIn [11]) and other, smaller ones, aim at linking professional communities (e.g. Shaping Tomorrow's Foresight Network [12] for the Foresight / Futures Studies community).

This phenomenon is currently undergoing intense research in social sciences [2, 13, 14] and in particular Human-Computer Interactions [15, 16, 17,18] Private sector companies are also investigating OSNs in order to learn about emerging lifestyles that may affect traditional business models. To our knowledge, the potential of OSNs for foresight activities has only recently been spotted [19] and we now introduce some of their essential features with this potential in view.

Participation in OSNs consists of joining as a member and interacting with other network members by sharing audio-visual content (e.g. Flickr [20], MySpace and YouTube [21]), contributing to forum discussions, exchanging views and ideas within communities of common interest (e.g. Orkut and Yahoo groups [22]), sharing sources of information (such as bookmarks in del.icio.us [23] and Digg [24]), collaborating towards a common goal (such as the encyclopaedia Wikipedia), and, last but not least, searching for and socialising with members with similar interests (most OSNs). To our knowledge, there is no commonly-accepted definition of "Online Social Network". However, they have two functionalities which make them stand out from other related online services:

1. Advanced tools for sharing digital objects (texts, pictures, music, videos, tags, bookmarks, etc.);
2. Advanced tools for communication and socialisation between members.

Amongst OSNs there is a great variety in the level of interaction and 'virtual socialisation'. These assets also determine the information that can be extracted from them. For the sake of simplicity (and without claiming to be rigorous in sociological terms), we can say that the more "modules" for cooperation there are, the more interactive the OSN is. Using this modular concept, we can classify online services in four main strands according to the level of interaction and socialisation they provide (Figure 1).

Online services in the first strand (e.g. Google Trends [25], Zeitgeist [26], Yahoo!Answers [27]) offer online access to massive amounts of information and knowledge, but contribute little to the "socialising dimension".

Online services in the second strand (e.g. Wikipedia [28], YouTube [29], del.icio.us, Digg) allow participants to share digital objects and collaborate towards a common goal. Even if they have a lot in common with OSNs, they do not foster socialisation between members as one of their main objectives or functionalities. Networks in the third strand (e.g. MySpace, FaceBook, Orkut, Friendster, LinkedIn, Flickr) are the core group of OSNs, strictly speaking the 'true' OSNs. They offer the possibility to join and contribute to a community of interest, i.e. ideally directed towards a coordinated action (co-action).

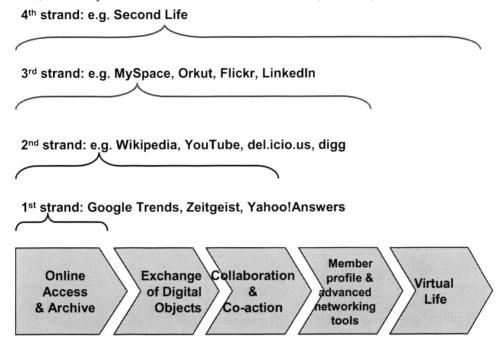

Figure 1. Level of interaction and socialisation for OSNs and other related online services.

Most (but not all) of them request would-be members to provide a set of personal or professional information when they register. This set of data forms the so-called "member profiles".

Finally, emerging technologies for 3D simulation of environments or persons (i.e. avatars) will allow for even richer socialisation and deeper immersion than current OSNs (e.g. virtual worlds, such as Second Life - see section 8). We believe that this is the emerging fourth strand of OSNs.

The third strand is the focus of our analysis, though we also draw examples from the other strands to illustrate our point.

A complementary type of classification of OSNs is also provided in a recent research paper [30] studying some leading OSNs. The paper shows that OSNs are 'scale free networks', i.e. they are organised around some central nodes and that they grow through the principle of «preferential attachment», i.e. the more a node has connections, the more chance it has to add new connections. The growth of such networks can follow two levers: either relying on its 'scale free network' structure, and inducing every new user, as a potential new network node, to bring all their 'real' connections into the virtual community, or by supporting the animation of the network with multiple and intuitive tools to interact with other users (since a connection node may not be an animator). Four criteria are identified for a typology of OSNs:

- The 'degree decentralisation', i.e. how far are interactions between users and profiles monitored and how "open" is the platform? (e.g. can users develop their own applications?);

- The 'number of different types of interactions allowed', i.e. is the network dedicated to a unique type of service (such as online matchmaking) or does it allow many more kinds of services?
- The 'type of identity', i.e. is the identity developed on the network close to the real identity of the user or is it a fantasised identity?
- The 'potential size of the network', i.e. what part of the Internet population might join the network? Niche vs mainstream network

Based on different dimensions in these variables four types of networks emerge (Table 1).

Table 1. Types of OSNs (developed on the basis of [14])

Type	Online communities	Business networks	Online matchmaking	Alumni networks
Goal	Socialising	Career / business opportunities	Soulmate	Getting back in touch
Examples	MySpace; FaceBook*; Orkut; Friendster; Skyrock	Linkedin; viadeo; XING	Match.com; meetic.fr	Copainsdavant.com; trombi.com
Degree of decentralisation	High	High	Low	High
Diversity of interactions	High	Medium	Medium	High
Distance from real identity	Medium to high	Low	Medium	Low
Potential size of network	High	High	Medium to low	High

* Facebook started as an alumni network and is now moving towards an online community.

INTRODUCTION TO FORESIGHT

From Pythia presiding over the Apollo Oracle at Delphi [31] to computer modelling, the drive to foresee the future is as ancient as humanity even if the modalities of attempting it are continuously evolving in the wake of new technologies, challenges and shifts in values.

In the 19th and early 20th century, science and technology were considered to be the major drivers of change. Looking into the future mostly consisted of anticipating forthcoming technological developments and most contributions were in the realm of "science-fiction".

Jules Verne and H.G. Wells [32] were amongst the most acclaimed authors in this field. During and after the Second World War, it became commonly accepted that the long-term status and welfare of a nation depend largely on its advancement in science and technology [33] and governments started to actively support technological developments. 'Technology Forecasting' drafted away from science fiction and became institutionalised, its focus remaining almost exclusively on scientific and technological issues. The technological experts and some professional futurologists were the main, if not the only, contributors to the field.

In the 1980s, "technology foresight" emerged when it became commonly accepted that the future is not pre-determined and that our current actions may contribute shaping it in a more desirable direction, in opposition to the deterministic future perspective of technology forecasting [34]. Gradually, foresight widened its scope when experience showed that the future was not only, or even primarily, determined by technological factors. For instance, the energy crises of the 1970's and 1980's were due to political rather than technological factors and had major economic and geopolitical consequences. However, many foresight experts did not see them coming. Accordingly, the emphasis shifted from technology and innovation alone towards trying to understand the trends and drivers of market(s) and economics in the so-called "second generation" foresight [17]. The key actors were then from both academia and industry, complemented by people able to bridge the gap between the two.

In the 1990s and 2000s, it became widely accepted that the development of new technologies was also highly conditioned by political, societal, psychological and cultural factors, such as concerns for the environment or the capability to understand and use technologies. Accordingly, this "third generation" of foresight acknowledged social determinants and behaviour as major drivers of change. In addition to the actors from academia and industry, social stakeholders such as non-profit organisations, consumer groups, lobbyists and government representatives have been regular participants in the foresight process on issues such as health, safety or environment [17].

Nowadays, Foresight, as defined by APEC Centre for Technological Foresight [35], refers to "*systematic attempts to look into the future of science, technology, society and the economy, and their interactions, in order to promote social, economic and environmental benefit*". Practitioners of foresight draw attention to three main concepts [36].

First, it is non-predictive, in as much as it does not pretend to be able to predict the future, which is unachievable in most cases, but to explore how the future might evolve depending on the actions of various players and decisions taken 'today'.

Second, it is oriented towards action. It not only analyses or contemplates the future but also aims at supporting actors in actively shaping it. The basis of the foresight approach is about opening minds to innovative possibilities. Building shared visions for the future is an effective way to shape it in a more desirable direction. Attempting to foresee possible future scenarios for the pure sake of knowledge, for instance the long-term future of the solar system, is not considered as Foresight [37].

Third, it is not limited to small expert groups but is participatory, and involves a wide range of stakeholders. As a consequence, foresight is not a restricted area for "specialists" but an open domain where individuals can share their views on a better future for our societies and ways to achieve it. Within the society, there is huge human potential and a lot of valuable tacit and codified knowledge in each individual that should be used as much as possible.

In the past decades, a large range of foresight methods, such as Delphi surveys, scenarios or roadmapping have been developed for a wide variety of settings [38,39,40] Popper proposed a typology of foresight methods in terms of the main type of knowledge source on which they are mainly based. Methods were classified under four sources of knowledge (evidence, expertise, creativity, and interaction). The sources of knowledge are not independent from one other; it is possible however, to use them to highlight the most representative features of each method.

The classification of the methods based on their main knowledge source is shown in Figure 2. The four sources of knowledge form the four corners of a diamond. The methods are placed inside this diamond. The degree of their proximity to the corners is an indication of the degree of dependence of each method on each of the four knowledge sources. For example, some methods may be 50% based on creativity, 20% on expertise, 20% on interaction and perhaps 10% on evidence. However, it has to be noted that it is subjective to evaluate the influence of these features or propose a generic weighting for each method given that practitioners use methods in different ways. Nonetheless, the "Foresight Diamond" provides a practical framework to help in the process of selecting methods based on their knowledge source.

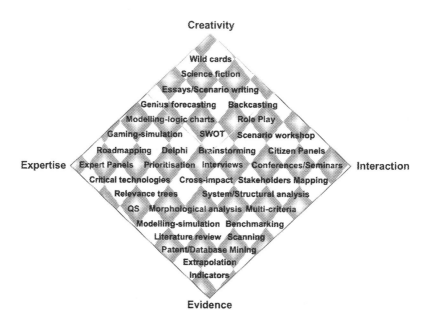

Figure 2. The "Foresight Diamond": Representation of a selected number of foresight methods regarding their contribution to four fundamental objectives, namely providing evidences, fostering creativity, building expertise on possible 'futures' and aligning interaction towards common visions (taken from 23 and slightly modified).

The bottom corner of the diamond includes methods related to collection and production of "evidence", i.e. available existing, or new, information and knowledge such as data and statistics. Indeed, it is necessary to study carefully the present state of a system before attempting to explain and/or anticipate a particular phenomenon.

The left corner of includes the methods that mainly utilise or build specific knowledge by exploiting the skills and 'expertise' of individuals in a particular area or subject. This expertise might be used to provide advice, make recommendations or support decision- or policy-making. The top corner refers to those methods that mainly serve to foster individual and collective "creativity", to generate novel assumptions, and thus enlarge the scope of possibilities for the future. These methods rely on the imagination, inventiveness and ingenuity of skilled individuals, such as science fiction writers, or the inspiration that emerges from groups of people involved for example in brainstorming or wild cards sessions.

And, finally the right corner involves methods facilitating the "interaction" between individuals and the alignment of individual contributions into collective and coherent views. Expertise gains from being brought together and challenged to articulate with other expertise (and also with the views of non-expert stakeholders). This is performed through 'bottom-up', participatory and inclusive activities.

Since the main foresight methods have been classified as depending upon four main sources of knowledge, they can also be considered as pursuing four main types of objectives, i.e. providing "evidence", building "expertise" on possible 'futures', fostering "creativity" and aligning "interaction" towards common visions.

Thus, Figure 2 also shows the main foresight methods in reference to the degree to which they pursue these four types of objectives. Building on this four-dimension taxonomy, this paper explores how OSNs could also be used within foresight research and practise to pursue the same types of objectives. We base our argument on the consideration that OSNs contribute in various ways to the four types of objectives. In the next section, we introduce OSNs as a social phenomenon and then analyse their potential from the perspective of foresight methodology.

OSNS FROM THE PERSPECTIVE OF FORESIGHT

Young people are the main and most active participants in OSNs [42] This population segment is particularly interesting as young people tend to adjust easier and quicker to cultural changes. They are what Rushkoff [43] calls "our evolutionary future". It is still debated whether early adopters (often young people in the case of ICT) will continue to be innovative as they age, or whether they will 'only' continue to use the technologies that they have learned in their youth. Moreover, it may be anticipated that some behavioural patterns observable amongst youngsters will also extend to other age brackets.

The proportion of the population socialising online is still marginal today but OSNs are increasingly becoming popular not only among the youth but across several age groups and classes of society.

A lot of data about users and their interaction is public. A good example of this kind of data is the member profiles, which is partially or completely visible to other members. The features of the profiling depend on the network and its target audience (e.g. young people, professional community). Member profiles can complement data gathered from other variables available through OSNs, such as date of discussion, list of friends or even simple tags.

Providing free access to (some) public data about users and their behaviour is not the only characteristics of OSNs. They are usually large networks offering a wide range of thematic topics as well as new modalities of large scale interaction. These characteristics make them worth considering for the purpose of foresight or futures studies [10].

OSNs can be regarded as (large) ensembles of (more or less) trusted people with specific profiles, personalities and knowledge who wish to share insights voluntarily. Starting from the assumption that it is possible to say the same for most foresight activities, the question arises of whether and how OSNs can contribute to exploring possible futures. The relevance of OSNs to foresight has recently been alluded to by Gartner Consulting [44]. For this purpose, we choose to examine OSNs and related online services in the perspective of the "Foresight Diamond" proposed by Popper [23], see section 2:

1. OSNs to provide evidences;
2. OSNs to build expertise;
3. OSNs to encourage creativity;
4. OSNs to foster interaction between individual thinking in order to build collective or "collaborative" intelligence.

These four dimensions are discussed in the following four sections. In the subsequent section, we discuss how "Metaverses" (or "metaphysical universes") will open new realms of creative applications for foresight and futures studies.

OSNs TO PROVIDE EVIDENCE

Within OSNs, there is a huge amount of information within publicly-available forums and archives which has not exploited for foresight purpose. Contributions to forums of one, or a few OSNs, can be scanned, data-mined, analysed, combined, extrapolated, using sophisticated ICT tools in order to derive meaningful knowledge. This can be thought of as a modernisation of the traditional approach of "environmental scanning" [19] to detect "weak signals" defined by Ansoff [45] as "*imprecise early indications about impending significant events*". Later on, this definition has been expanded to accommodate additional characteristics, such as "*new, surprising, uncertain, irrational, not credible, difficult to track down, related to a substantial time lag before maturing and becoming mainstream*" [46].

The characteristics of the underlying knowledge and the way it can be extracted depend on the size of the network. Within a small focus network (e.g. alumni networks, professional networks focussed on a specific community) members will fill out their profiles and their professional affiliations, provide information on their other centres of interest and give indication on their level of expertise in a given subject. Moreover, members do not want to be seen making useless contributions in view of their community, thereby ensuring a high level of discussions in the forums. The use of a well-bordered OSN to detect social changes has already been explored by Adamic *et al* [47] in their study of the student community at Stanford University, called 'Club Nexus'. They have found correlations between Nexus members' real lives - their personalities, tastes, hobbies, academic majors and their genders - and their online profiles.

In large OSNs (e.g. MySpace, Orkut, Friendster, Facebook), information can be extracted and knowledge generated from millions of seemingly uncorrelated bits and pieces. Emerging changes in (particularly social) behaviour can be detected on the basis of thoughts, opinions and information provided in the multitude of discussion forums. Moreover, it is possible to relate the contributions and interests of a specific member or a specific group of members to his/her/their profile(s) and to his/her/their additional contributions to other forums. Extrapolation of such profile data can provide additional insights about online behaviour [48] As in the physical world, the value of profiling in the virtual world is also related to the level of detail and veracity of the subject, being aware that there is a wide range of profile options in some OSNs. This extraction is not straightforward as the information in the profile is often not reliable with many fake profiles, changing nicknames and non-human accounts. There is little possible cross-checked with real life information. Fake profiles are commonly known as 'fakesters', a contraction of the term 'fake' and the OSN 'Friendster'. This refers to a type of account pertaining to bands, movies, celebrities or brands etc. However, even if these creative accounts may seem irrelevant on the surface, they are nevertheless significant, as they have enabled networking between friends and connected new people with similar interests through an intermediary [49] in a way not unlike communities of interest [13].

Adaptive gathering and effective analysis of unstructured information is a difficult task. In view of the large number of OSNs and their rapid development, one important issue is how to scan networks in order to extract the useful signal (emerging trends) from the very noisy background (the bulk of poor contributions). Detecting weak signals of changes in social behaviour requires effective and thorough information gathering, and a solid methodology for its analysis. As mentioned above, the analysis cannot be a one-size-fits-all one but needs to be adapted to the specific research question under investigation. Intuitive selection criteria include the size and level of activity of the forums and the area of activity. A fraction of these querying and matching tasks will probably remain 'manual' and –in view of how time consuming it is to collect useful information- a previous screening of suitable sources may be unavoidable. Another issue is how to ensure that results are statistically significant and provide a non-biased representation of the online population. In view of the specificity, this needs to be discussed case by case.

Also it is would be particularly fruitful to compare analyses of archived snapshots of the network at regular intervals to pinpoint signals of emerging trends across time.

It will soon be possible to cross information from various OSNs with "Social Aggregators" such as Snag [50], ProfileLinker [51] and many others, which address the problem of having to set up OSNs on multiple platforms by integrating data from various OSNs into a single web application [52].

Social scientists at Indiana, Northwestern, Pennsylvania State, Tufts, the University of Texas and other institutions are mining Facebook to test traditional theories in their fields about relationships, identity, self-esteem, popularity, collective action, 'race' and political engagement [53]. Private sector companies are also investigating OSNs in order to learn about emerging lifestyles that may affect traditional business models and for targeting advertisement. The figures speak for themselves: MySpace was sold in July 2005 to the press magnate Rupert Murdoch for a price corresponding approximately to 35 US$ per user profile for ($580M) and YouTube to Google Inc in October 2006 for $1.65 billion in a stock-for-stock transaction.

We are not aware of public search engine for automated gathering, querying, retrieving, matching and displaying of information across OSNs, even though similar engines might emerge in the future from the already existing systems to analyse unstructured data, such as WebFountain developed by IBM [54].

In [13], we address the changes in communication behaviour observed through 'Google Trends' and 'Google Zeitgeist'. Although they are not OSNs in the sense discussed in Section 1 (see also Figure 1), 'Google Trends', 'Google Zeitgeist' and Internet traffic measurements, like the ranking offered by the company Alexia [55] are valuable online services for detecting and observing changes in search behaviour. They do help us to understand what OSNs could offer by analogy. 'Google Trends' presents the volumes of search queries of specific keywords over time and, where possible, relates the ups and downs in the curve with specific news articles. 'Google Zeitgeist' collects the most frequent search queries on the web and presents a hit-list of keywords by time periods (weekly, monthly and yearly), by topic and by country. If Internet queries can also be regarded as an indicator in the search to address user needs, concerns and topics, then this hit-list reflects the state-of-the-art for a given date and country. (It may be assumed that, for established inquiries, users tend to access the URL directly, while for new topics and inquiries, search engines are more often used as an entry point to find relevant links). Based upon this assumption, the top ranking keywords in the Google Zeitgeist lists can be regarded as an indicator of the "temperature of hot topics" over time amongst online users, and a selection by countries may offer information about cultural differences.

Finally, artificial environments, such as Second Life (see § 8) also provide obvious insights to changes in social behaviour.

OSNs to Build Expertise

OSNs enable members from all over the world, whether they are knowledgeable or only interested in a specific subject, to enter into contact and share information or knowledge. It is noteworthy that in certain fields, foresight is one example; there might be only a handful of knowledgeable experts within small- or medium-size countries. OSNs allow the creation of larger pools of people around common interests. In comparison with the web itself, OSNs have an interactive dimension in the sense that members can connect to each other to clarify, discuss or ask for complementary information.

In large OSNs (e.g. MySpace, Orkut, Friendster, Facebook), there are millions of potential contributors in discussion on any potential subject. Certainly, many of them are novices in the issues at stake, capable only to offer non-informed opinions. Even if this were the case, the number of members is so large that there remain many who are both knowledgeable on the issues under discussion and willing to contribute. Actually, the simple fact that many contribute voluntarily can be taken as an indication that they are either knowledgeable or highly interested in the topic under discussion. Therefore, there exist many interesting constructive contributions within the bulk of poor contributions; the important thing is the selection process to extract them. Irrespective of the contributor profiles, individual contributions are generally contested, which generates a dynamic debate, as in the case with Wikipedia. The more dynamic a network is, the more likely it is to attract new members.

Social bookmark managers such as 'del.icio.us' and 'digg' can also inform of emerging popular topics on the web, which, can be used to reach a better understanding of behavioural trends. First, they have a huge number of subscribers and feature both individual and collective behaviour. People can view the lists of links collected by others, and subscribe to the links they find interesting, thus, initiating a viral effect. At the same time, it is, in a way, self-regulating as any user can see the links that others have collected and who else bookmarked a specific site. Second, most of them offer a dynamic list of most popular tags which is based on analysing a specific tag, related tags and associations between tags. Emerging trends can be observed by monitoring changes of popularity over time. Third, the selection of tags can provide information about online semantic relations. More detailed information about the meaning associated with online tags can be obtained by a systematic analysis of tags, the number and content of out-links. Within [13], the tag 'environment' is first assessed for del.icio.us and then for YouTube on one particular day. The outcomes show that even if both systems are fuelling public outcry over environmental concerns and the responsibilities of the decision-makers, they deal with the same topic in completely different ways that reflect their specific culture. While in del.icio.us members aim to share critical information from trusted sources, in YouTube criticisms are usually expressed through irony and 'fun' without necessarily going into the complexity of the issues. We also describe the OSN '43things.com' [56] which provides an interactive social platform, where users can communicate what they would like to do in the near future [13]. The visions that (mostly young) members have of their own future is obviously an important indication of the way they will attempt to shape their life, and this is thereby relevant in the context of foresight.

Finally, we look at Care2 [57] [13]. Care2 is a OSN which aims to put activists from around the world in contact. Its stated mission is "to help people make the world a better place by connecting them with the individuals, organisations and responsible businesses making an impact." In this context, emerging changes of social behaviour can be highlighted by monitoring the major themes of discussion. For instance, a list called the "most favourite causes", a hit-list of areas to fight for, and another called "what scares us most", comprising the topics of fear and concerns, are provided. Some of these fears appears to be deeply rooted in the human nature and therefore relatively permanent, such as "losing the ones I love". Others, however, infer changing social patterns, such as "being alone" for smaller and fragmented families and single-person households. Again others, like "religious extremism", may be a sign of spontaneous reaction to current events. What matters more in this example may not be the absolute membership levels but how they will evolve over time.

OSNs TO ENCOURAGE CREATIVITY

Creativity has been conceptualised as:

1. The individual personality traits that facilitate the generation of new ideas,
2. The process of generating new ideas,
3. The outcomes of creative processes, and
4. The environments conducive to new ideas and behaviour (Rhodes, 1961 as cited in [58]).

These perspectives led to multiple definitions of creativity. For example creativity is the capacity to generate new and valuable ideas for products, services, processes and procedures, or the ability to produce work that is both novel (i.e., original) and appropriated (i.e., useful), or the set of qualities of products or responses that are judged to be creative by appropriate observers. (Martins and Terblanche (2003); Sternberg (1999); Amabile (1996) as cited in [34])

Irrespective of the point of view (person, product, process) the concept of creativity is approached, two specific levels of creativity can be identified: the radical or revolutionary and the adaptive and confirmatory [34] similar to Abernathy's notions of 'radical' and 'incremental' innovations. Radical or revolutionary creativity is associated with Kirton's 'innovative cognitive style' of solving problems, or Wertheimer's 'productive thinking' consisting of the combination and integration of different known principles and ideas into a new combination or with de Bono's 'divergent-lateral' thinking. Adaptive and confirmatory creativity, on the other hand, can be linked to Kirton's 'adaptive cognitive style', or Wertheimer's 'reproductive thinking' applying a known solution to a new situation or with de Bono's 'convergent-vertical' thinking characterising problem-solving along known lines and according to established principles. (Kirton 1987; Wertheimer 1945; DeBono 1971 as cited in [59])

Organisational conditions that restrict free and open communications (e.g. rigid bureaucratic rules and instructions, 'holy' hierarchies and detail controlling supervision) keep creativity down. Conservative values, strategies and policies, that support the conventional, also block creativity. Time pressure, heavy work-load and stress reactions hamper the incubation phases and subconscious mental activities, which are elements of the creative problem-solving process, and need time for retreat, reflection and relaxation. Risk-taking and anxiety on the other hand are ingredients of creative acts. Thus, in environments where creative initiatives are met with suspicion, defensiveness and aggression, the fear of failure holds creativity back. However, it is not that in strict, structured cultures and praxis no creativity is developed. It is the adaptive level of creativity that is more likely to develop in such environments than the radical one. Radical creativity is enabled in environments with looser structures, more freedom, higher risk-inclination, and debating, dynamic and playful atmosphere [35].

Walton [60] suggests that this distinction (adaptive vs. radical creativity) is broadly equivalent to Heinzen's (1994) categories of creativity: reactive and proactive. Reactive creativity occurs as a function of some external stimulus, and is often goal oriented. Proactive creativity, on the other hand, is internally driven and produces results that may be less directly related to the solution of a particular problem.

The internal desire to be creative has often been highlighted in the literature as a major factor for individual and in this respect collective creativity too. Amabile [61] notes that for enhancing creativity in businesses three components are needed: technical, procedural and intellectual knowledge and expertise; creative thinking skills (how flexibly and imaginatively people approach problems) and motivation (especially intrinsic motivation, i.e. internal desire to do something based on interests and passions).

As other prerequisites for creative performance the role of interaction with other individuals especially from diverse backgrounds has also been noted. For example Fischer [62] characteristically notes that *"Although creative individuals are often thought of as working in isolation, the role of interaction and collaboration with other individuals is critical to creativity... Distances (across spatial, temporal, and technological dimensions) and diversity (bringing stakeholders together from different cultures) are important sources for social creativity."*

Alves, et. al. [34] agree that such environments, characterised by diversity of actors and competences, coherence respecting the integration of complementary activities and interactivity, i.e. strong cooperation relationships, maximise the benefits of cooperation and guarantee that learning effects and levels of inventiveness are enhanced due to higher cultural, technical and knowledge differences between the actors involved. They reinforce creative competences and allow for rich combinations of otherwise disconnected pool of ideas. In the cases they study to draw these conclusions they also note that trust and confidence, ensured by excluding direct competition and agreeing upon confidentiality and intellectual property codes, allowed for open and frank discussions.

Referring to creativity and social network theory, Shalley, et. al. [63] highlight too the importance of 'weaker ties' (i.e., more distant relationships, such as acquaintances or distant colleagues) asserting that they might be more beneficial for creativity than stronger ties (i.e., good friends or close relationships) because they enable communication of novel, non-redundant information from diverse social circles. Additionally they note that individuals' positions in their own network, as well as the connections they have outside their network could influence their creativity.

In studying the effects of personal and contextual characteristics on creativity Shalley, et. al. [39] conclude that several contextual characteristics have consistent, significant effects on individuals' creativity and that the direction of these effects is in line with the intrinsic motivation perspective. For example, evidence is rather clear to support that individuals tend to exhibit high creativity when their jobs are complex (and thus they enjoy more autonomy as well as responsibility), their supervisors engage in supportive, non-controlling behaviours, their work is evaluated in a developmental, non-judgmental fashion, and their work setting's configuration restricts unwanted intrusions.

Understanding the factors affecting the levels of individual, group and organisational creativity is a complex research area. Researchers agree that this area of research is still quite young. Additional characteristics should be studied like intrinsic motivation as a mediator, individual mood states, self-efficacy and creative role identity, existence of creative role models, the role of creative processes, creativity in international contexts, different types of creativity, the measurement of creativity, team creativity as well as social networks [39]. The gaps regarding the factors that influence the many dimensions of creativity are still many, especially regarding how creativity is affected by sociological and social-psychological factors such as social norms and group influences, leaving many questions unanswered regarding the environment / individual interface [39].

Nevertheless, certain arguments can be supported asserting that OSNs can foster creativity. In examining how OSNs may affect creativity we have to know the characteristics of the OSNs and how these evolve.

Although OSNs do have a form of internal organisation and procedures (see § 1), they can by no means be regarded as strict structures nurturing conservative environments and supporting conventional values. Instead they are quite loose structures allowing full freedom, open communication under dynamic and playful atmosphere. They lack any type of supervision that can control behaviours, or any type of judgement that may restrict free expression and creativity. In this respect they can be considered as environments fostering creativity. Of course this freedom also allows any improper behaviour between members, which may make some leave, but this is a trade-off in networks of this type.

Secondly, the birth and growth of online OSNs is based on the internal desire of people to do something based on interests and passions. This is the same intrinsic motivation where individual creativity also depends on.

Thirdly, OSNs bring together individuals from different background, views, perspectives, competences and experiences. These diversities and 'weaker ties' are linked under the mutual desire to exchange views, data, and information and interact with, learn about and from each other. This is the type of environment where creative thoughts and ideas can be born.

Even though other factors enhancing creativity such as strong relations, trust and confidence are less evident, an argument supporting that such networks indeed foster creativity of their members does not at all seem far-fetched. The interactivity of their functions, and the way they are appropriated by users have enabled new social processes, where communication was previously not even thinkable. They have enabled millions of individuals to express and diffuse their creativity into new realms, from sharing bookmarks on webpages on del.icio.us, publishing photos on Flickr, to sharing videos on YouTube and MySpace [13]. They can access to a large audience and get recognition and feedback. Further interaction takes place through voting, discussion, and providing additional audio-visual materials that contrast or complement each other.

In addition, OSNs expand the concept of community by allowing people from different corners of the world to communicate and reciprocate in real-time at a negligible cost through the use of multimodal channels, including text and audio-visual material. In [13], we illustrate the popularity of Podcasting which was declared by the editors of the New Oxford American Dictionary 'Word of the Year 2005' and define as "a digital recording of a radio broadcast or similar programme, made available on the Internet for downloading to a personal audio player" [64]. We also present YouTube and MySpace as examples of how OSNs can foster creativity. With regard to foresight, these new tools become interesting whenever this creativity is used to generate plausible new assumptions about the future and thus enlarging the range of possibilities for the future. In the near future, it is not far-fetched to imagine that amongst millions of OSN members, there are large numbers of people who will simulate and experience future scenarios in virtual reality (see Second Life § 8).

OSNs to Foster Interaction

Collective and "collaborative" intelligence emerges from the collaboration (and competition) of many individuals, where the resulting intelligence is larger than the sum of individual contributions [65]

Building collective intelligence is a necessary step previous to decision making in order to reduce individual cognitive bias and to ensure that the decision will be accepted and supported by a wide majority once it is taken. The deliberations in the fifth century BC Athens, which established the historic prototype of a democratic society under Pericles' leadership, can also be seen as a prototype of building collective intelligence out of individual contributions and interest, what George Pór defines as "the capacity of a human community to evolve toward higher order complexity thought, problem-solving and integration through collaboration and innovation" [66].

Similarly and as explained in Section 2 one of the main functions of foresight is to facilitate the interaction between individuals in order to contribute to the alignment of different perspectives into a coherent and solid collective view. Many foresight methods imply the organisation of face-to-face encounters between people (e.g. expert groups, brainstorming) in a same geographical place, in close (physical) contact. The number of participants in these events is limited because, besides logistics and budgetary reasons, the quality of interaction quickly decreases as the number of people increases.

The overall goal is thus to improve communication and understanding between people with new information-sharing, knowledge-transmission and content-exchange practices. Indeed, the process of collective thinking has not changed much over the last 2500 years; from let's say the deliberations on the Acropolis in the fifth century BC to the board of a 21st century corporate. It is still based on the talking of single individuals (nowadays with the help of PowerPoint presentations), each at his/her turn, and the others listening.

It is well-known that the rate of information transfer between the speaker and a few listeners is very poor. One is talking, a few are listening attentively, and others have lost attention, or are busy thinking to previous contributions, future ones that they are going to make or side issues. Moreover, no speaker is able to present the whole context in which his/her statement is embedded. Therefore statements are often misunderstood by the listeners. Researchers have confirmed this ineffectiveness of the communication process by tape-recording conversations among people. A thorough analysis of this material has demonstrated how little we understand each other in such conversations, mostly talking at cross purposes [67]

As illustrated in various examples by Tapsott and Williams [68], the new low cost collaborative infrastructures (from free-Internet telephony to open-source software) have created new modes of communication, innovation and value creation. This is based on 'peer production', where self-organised masses of individuals and firms collaborate openly to drive innovation and growth [44]. Tools contributing to collective intellectual content are high on research agendas. In fact, the consultancy firm Gartner has identified collective intelligence as one of the technologies with the greatest potential for business over the next 10 years [26]. These applications include technologies such as code and documents, through individuals working together with no centralised authority and it is seen as "a more cost-efficient way of producing content, metadata, software and certain services". These approaches are expected to become mainstream in 5 to 10 years.

In similar lines Benkler [69] who was first to coin the term 'commons-based peer production', argues that a 'networked information economy' emerges, displacing the industrial information economy that typified information production from about the second half of the nineteenth century and throughout the twentieth century. This new stage of the information economy is characterised by decentralised individual action, specifically, new and important cooperative and coordinated action carried out through radically distributed, non-market mechanisms that do not depend on proprietary strategies. The rise of the networked, computer-mediated communications environment has helped remove the physical constraints on effective information production. This is what has made human creativity and the economics of information itself the core structuring facts in the new networked information economy and lead to the rise of effective, large-scale cooperative efforts - peer production of information, knowledge, and culture.

Benkler [45] notes that collaboration in peer-production is usually maintained by some combination of technical architecture, rules, and a technically backed hierarchy of meritocratic respect, but also on social norms, and to a great extent, on the mutual recognition that it is to everybody's advantage to have someone overlay a peer-review system with some leadership. Private and social motivations appear that are different from the dominant driver of profit-making. Motivations such as getting pleasure in sharing, acknowledgement by peers, enhancement of reputation, positive network effects due to increased diffusion and the possibility that others may further improve or suggest improvements to the innovation to mutual benefit outweigh potential losses or risks of loss. Even a modest contributor, or even a non contributor but who could potentially become one, can feel some ownership for the common undertaking. It becomes more difficult to criticize the shortcomings when everybody is provided with the tools to correct those and improve the product. Also, it becomes less likely than a competitive undertaking will be launched in parallel thereby misspending human and financial resources.

Supporting von Hippel's notion of 'user-centred innovation' [70] Benkler adds that the other quite basic change brought by the emergence of 'social production', from the perspective of businesses, is a change in taste. Active users require and value new and different things than passive consumers did. Consumers are changing into users - more active and productive than the consumers of the industrial information economy.

In [13], we provided various examples of systems to mediate and merge individual contributions into collective intelligence, first with Yahoo! Answers which operates as a large-scale online brainstorming tool by asking forward-looking questions and seeking creative responses to them, and then with Wikipedia, a prime example of a collaborative intelligence tool to construct an online not-for-profit encyclopaedia. Wikipedia provides space for creative contributions from any users who are willing to contribute. The value of text is based on the fact that it can be contested and edited by different users. The success of this undertaking shows that a multitude of individual contributions, from both professionals and amateurs, can coagulate into a successful collaborative endeavour. Apparently, advanced software tools are used in industry to design collaboratively complex systems such as cars or planes.

It would be a real breakthrough in the organisation of smaller scale one-to-few or few-to-few collective processes if it were possible for listeners to engage, react or contribute in real time. Certainly some recent developments go in the direction of enhancing the efficiency of foresight methods through the use of software, like mind-mapping [71], electronic voting by participants [22] or the online (so called 'real time') Delphi, which enables the participation of more brains in studies [72]

In this sense, OSNs appear as a promising direction of investigation in the sense that they can allow for the (partial) replacement of some of these face-to-face encounters by more efficient online processes which would also be open to more participants. The major drawbacks of online interactions are the lack of visual contact, spontaneous reaction, non-verbal communication and other physical interactions. This reduces the richness of communication and group dynamics and these shortcomings can be only partially compensated for by good online moderation.

The question arises whether OSN advanced communication modalities can counter-balance for the lack of face-to-face interactions by integrating richer contributions from a larger number of participants. At the same time this number of participants has to remain below 150 what seems to be an optimal size for human groups [73]. Most OSNs operate without a moderator, participants are free to participate and interaction is not limited by time and space boundaries. At the same time, such characteristics facilitate more interactive, versatile and thoroughly thought-through contributions.

The open structure of OSNs allows researchers and respondents to interact and participate on an equal level, whether they are experts or novices, as long as they share similar interests and motivation (joining thematic communities). In the near future OSNs –if well managed-could operate as a large-scale method for online discussion, brainstorming, and questions on forward-looking topics, as a test-bed for concepts, ideas, assumptions or scenarios.

However, how valuable the contribution of OSNs to collective intelligence might be, they cannot pretend to solve all the problems related to lack of attention or misunderstanding. Actually they may even make some of these worse by encouraging multi-tasking. And certainly they also run the risk that –similar to real life forums or online discussions- they become free-running with little possibility to steer discussions towards a precise objective.

Systems which are able to scan, select, aggregate and display news in a user-friendly way according to the profile of the users are already appearing: Google News [74], LeapTag [75] which can plug in into Facebook, Digg which combines social bookmarking, blogging, and syndication with a form of non-hierarchical, democratic editorial control [76] Yahoo Pipes [77] which aggregates, manipulates, and 'mashups' content from around the web using a process that can be specified or Twine [78], which has been described as the "the first mainstream Semantic Web application".

These systems rely on the input (e.g. tagging, bookmarking, voting, selecting as favourite) of many contributors. When the underlying technology will further improve, it is expected that similar systems can become relevant even with a smaller underlying basis or contributors. Therefore it will be possible to adapt them to the need of smaller professional communities.

In the near future, 'intelligent' systems will be able to synthesise and merge a large number of information and contributions into an operational summary which will be tailored according to user specifications, tastes, knowledge of the topics and availability, and which will continuously adapt itself according to the feedback received from the user.

Further technological innovations within the OSN spectrum, such as using avatars or three-dimensional animations, anticipate innovative methods of conducting foresight exercises (see Second Life § 8). The ability to brainstorm, discuss and collaborate through a real-time interactive platform already opens up new modes of foresight practice. As Glenn stated "future participatory systems could include global cyber games with millions of participants to create policy." [79].

OUTLOOK FOR THE FUTURE: METAVERSES AND SECOND LIFE

Second Life (SL) presents a three dimensional (3D) virtual world, also called "Massively Multiplayer Online Role-Playing Games" (MMORPG) or "Metaverse" (for "metaphysical universe"), which is developed and owned by its users, referred to as 'residents'. SL has around 100,000 active users (February, 2007) interacting together online. The inhabitants of SL emulate online real life activities such as attending classes, setting up businesses, going to parties, forming relationships, creating associations and self-help groups. The properties of the virtual space are continuously evolving according to the advanced level of social networking taking place between the residents. Notwithstanding the open participatory environment, the monetary variable plays a major role and without it, residents are not able to do much.

Unlike 3D games, SL does not have points, winners or gaming strategies, but focuses on the interactive life of its residents. Thus, SL is more like a virtual-world OSN than a cyber game (see Figure 1). Many organisations, such as companies, NGOs or political parties already use SL as a platform for organising, training and developing their activities. SL is one of the 3D virtual worlds which are currently been built in a large variety of context. The company IBM is developing a Metaverse to help its employees collaborate. It has a large variety of landscapes with a waterfall, an amphitheatre and an underground cave. It is not well known how extensively it will be used throughout the company [80]. The US space agency NASA is exploring the possibility of developing a Metaverse aimed at students to "simulate real engineering and science missions" and "help find the next generation of scientists and engineers needed to fulfil its vision for space exploration. (...)The MMO will foster career exploration opportunities in a much deeper way than reading alone would permit and at a fraction of the time and cost of an internship program" [81].

We believe that technological developments in this area, where motional avatars can communicate and interact in a real-time, 3D, self-created, *quasi* science-fiction world can be of significant use for foresight practitioners. For the first time, these innovative tools would allow foresight practitioners to make use of real-time interactions, 3D environments and motional avatars. In these conditions, fictitious future scenarios can be simulated, and virtually co-habited by the residents under research.

These virtual spaces could act as test-beds to foresee the potential and drawbacks of emerging technologies or to anticipate the effects of potential business developments or policies, thereby better preparing them. They can be used to make key players or lay people aware of possible drastic consequences of global warming or military confrontations. It could be possible to observe the behaviour of avatars in a free-market fast-growing economy with little control or on the contrary in a centralised economy with many regulations, or to simulate the consequences in terms or law and order of the loosening or the hardening of certain laws. It is worth adding that these scenarios could take place either in completely fictitious environments or within the digitalisation of real environments as many cities are being digitalised, sometimes even at different times in history [82, 83]

Innovative modes of foresight practice can be developed and combined based on data derived from 3D real-time interactive environments, forums and events organised within the same *futuristic* environments. It always has to be kept in mind however, that the outcome of such simulations is highly dependent on the profiles of the people participating and also the rules of the games which may differ from those in different socio-economic, political and institutional environments.

CONCLUSIONS

In this chapter, the authors suggest that OSNs could be used as complementary methods in futures studies or foresight. The foresight toolbox is already rich of many different methods

[20] and new ones will continue to be added in the future. As the phenomenon and the usage of OSNs are relatively unexplored for foresight purposes, our aim is to trigger further thinking about their potential and limitations, rather than to present firm conclusions. In our view, OSNs can contribute to foresight in four ways.

First, OSNs can provide evidence, data and information, by analysing recorded exchanges of information and thoughts amongst participants in discussion forums and cross-crossing it with other available information, such as member profiles, behaviour or networking patterns. This way, they can be used as expert systems to detect and monitor emerging changes in social behaviour. How far this is applicable in practice is unclear. For instance, participant profiles are not always reliable but rather target to transmit idealised or fake images of their owners. Anyhow, the changes in communication patterns and the increasing trust within collective communities presented in the text are *per se* examples of emerging social trends that can be detected by OSNs. As similar approaches are already being used in market research for commercial purposes, we are hopeful that some of these experiences could be transferred to foresight.

Second, OSNs can build expertise by facilitating the sharing of information and knowledge between members from all over the world, on any possible issues. Discussion forums offer a dialogue platform for interested and both knowledgeable and non-knowledgeable members. The former join in and are eager to express their 'expert' views, the latter are being most receptive to the 'expert' knowledge offered at no cost. Apart from existing knowledge diffusion, OSNs contribute to new knowledge production by being valuable sources of information about which are the topics and issues that concern people the most for the future and what people discuss about them (like the outcomes under for example 43things, or Care2).

Third, OSNs provide the proper environment to encourage creativity. Moreover, they offer functionalities that directly foster creation of any kind through the unprecedented modalities of communication and interaction on a large scale. The possibility of sharing multi-media material produced by their members is already a trigger for disseminating creative works by individuals and, in certain cases, getting it recognised worldwide. In the future, we expect that even more virtual environments, such as participant-enabled avatars in 3D environments like Second Life, could be used for creating foresight scenarios in a virtual world or carrying out large-scale simulations, with many profiles at a reasonable cost.

And last, OSNs are means of building collective or "collaborative" intelligence out of individual thinking for a whole range of possible future goals. The construction of the Wikipedia encyclopaedia shows that dispersed individuals and inhomogeneous groups can cooperate even without knowing each other, via an *ad hoc* ICT system, and produce excellent results. They do this by auto-regulation of internal thematic controversies with a minimum of operational management. We think that –if a similar suitable framework can be established-OSNs may also provide a meeting point for foresight practitioners and interested individuals. However, designing and running such a tool is not straightforward. In traditional foresight settings, the coordinator has a reasonable influence and can steer the discussion towards a given objective. In a less structured and non-hierarchical setting, like OSNs, the role of the coordinator is far less influential. The risk is that discussions may drift away without leading to useable results.

A tested research framework, which could be used as an analytical tool to conduct foresight, is desirable. In this respect, various research methods and alternative approaches should be taken into consideration to get the best possible results. This includes approaches to extracting relevant information, the creation of specific foresight forums, and best-practices on how to operate these. One example of a practicality that should be dealt with is the fact that foresight activities are generally defined in time (they have both a starting and an ending), while OSNs have open-ended character, with people sometimes responding with a long time delay. This is not a problem per se, but poses constraints in setting the boundaries of the research framework.

Another question of practice is how far the use of OSNs may infringe the privacy of their members. Although information on the web is voluntarily provided and publicly accessible, there might be ethical concerns about monitoring people's behaviour without their consent (e.g. participants may not want to be part of a research project). Another difficulty is that archived OSN information is largely unstructured; the underlying data is non-codified and difficult to extract. In addition, textual data may be incoherent, contradictory or incomplete, making automatic screening of information unviable.

Using OSNs for foresight purposes may not be easy, but we are confident that its potential makes it worth the effort. The booming world of OSNs make it easier to be imaginative than in the times of Jules Verne or H.G. Wells.

REFERENCES

[1]   A.-L Barabási, *"Linked: The new science of networks"*, Perseus, Cambridge, 2002.

[2]   L. Garton, C. Haythornthwaite & B. Wellman, *"Studying Online Social Networks"*, Journal of Computer-Mediated Communication, 3 (1) (1997).

[3]   W. Sack), What Does a Very Large-Scale Conversation Look Like?, in: N. Wardrop-Fruin & P. Harrigan (Eds), First person: New media as story, performance and game, MIT Press, Cambridge, 2004.

[4]   B. Wellman, *"Computer networks as social networks"*, Computers and Science (293) (2001) 2031 – 2034, p. 2031.

[5]   M. Castells, *"The rise of the network society"*, Blackwell Publishers, Malden MA, 1996.

[6]   Wikipedia, List of social networking websites, http://en.wikipedia.org/wiki/List_of_ social_networking_websites

[7]   S. Farnham, S.U. Kelly, W. Portnoy, J.I.K.Schwards, Wallop: Designing social software for co-located social networks, In Proc. HICSS-37, 2004, Hawaii (http://csdl2.computer.org/comp/proceedings/hicss/2004/2056/04/205640107a.pdf).

[8]   D. Boyd, Identity production in a networked culture: Why youth heart MySpace", American Association for the Advancement of Science, St. Louis, MO, February 19, 2006, (http://www.danah.org/papers/AAAS2006.html).

[9]   T. Erickson & W. A. Kellogg, Social translucence: An approach to designing systems that support social processes, ACM Transactions on Computer-Human Interaction, 7,1, (2000) (http://delivery.acm.org/10.1145/350000/345004/p59-erickson.pdf?key1 =345004&key2=4794773611&coll=&dl=ACM&CFID=15151515&CFTOKEN=61846 18).

[10] F.B. Viegas, M. Wattenberg & K. Dave, Studying cooperation and conflict between authors with history flow visualisations, In Proc CHI 2004, April 24-29, 2004, Vienna, Austria, (http://alumni.media.mit.edu/~fviegas/papers/history_flow.pdf).

[11] S. L. Bryant, A. Forte & A. Bruckmam, Becoming Wikipedian: Transformation of participation in a collaborative online encyclopaedia, in Proc, *GROUP'05,* November 6-9, 2005, Florida, USA, (http://www.cc.gt.atl.ga.us/grads/f/Andrea.Forte/BryantForte BruckBecomingWikipedian.pdf)

[12] C. Marlow, M. Naaman, d. Boyd & M. DavisHT06, Tagging Paper, Taxonomy, Flickr, Academic Article, 2006, (http://delivery.acm.org/10.1145/1150000/1149949/p31-marlow.pdf?key1=1149949&key2=9816773611&coll=&dl=ACM&CFID=15151515& CFTOKEN=6184618)

[13] R. Cachia, R. Compañó, O. Da Costa, Grasping the Potential of Online Social Networks for Foresight, Technological Forecasting & Social Change 74, 1179-1203, September 2007,
(http://forlearn.jrc.es/guide/5_running/documents/TFSC%20Cachia%20Compano%20 DaCosta%20Foresight%20SN.pdf).

[14] faberNovel Consulting (2007), "Social Network websites: best practices from leading services", Research Paper (http://www.fabernovel.com/news/research-paper-social-network-websites/, last accessed 27 February 2008)

[15] Wikipedia, Pythia, http://en.wikipedia.org/wiki/Pythia

[16] V. Bush, Science the endless frontier, A report to the President by Vannevar Bush, Director of the Office of Scientific Research and Development, July 1945, (http://www.nsf.gov/about/history/vbush1945.htm).

[17] L. Georghiou, Third generation foresight: Integrating the socio-economic dimension, In the proceedings of International Conference on Technology Foresight, Science and Technology Foresight Center of NISTEP, Japan, 2001 (http://www.nistep.go.jp/ achiev/ftx/eng/mat077e/html/mat077oe.html).

[18] Asia-Pacific Economic Cooperation - Centre for Technological Foresight, (http://www.apecforesight.org/)

[19] FOR-LEARN Online Foresight Guide, Characteristics of Foresight, (http://forlearn .jrc.es/guide/1_why-foresight/characteristics.htm).

[20] FORLEARN Online Foresight Guide, Description of main methods, (http://forlearn.jrc.es/guide/4_methods/methods.htm).

[21] J.C. Glenn & T. J. Gordon (eds), Futures research methodology, AC/UNU Millennium Project, Version 2.0, 2003, (http://www.acunu.org/millennium/FRM-v2.html).

[22] M. Rader & A.L. Porter, "FTA assumptions: Methods and approaches in the context of achieving outcomes", Second International Seville Seminar on Future-Oriented Technology Analysis, 28-29 September 2006, (http://forera.jrc.es/documents/ papers/anchor/FTA-Paper%201-Porter%20%20RaderFinalPaperV4-aug5.pdf).

[23] R. Popper, "Foresight Methodology", in Georghiou, L., Cassingena H., J., Keenan, M., Miles, I., and Popper, R., "The Handbook of Technology Foresight: Concepts and Practice", Edward Elgar, UK, 2008.

[24] A. Lenhart & M. Madden, Teen content creators and consumers, Pew Internet & American Life Project, 2005, (http://www.pewinternet.org/pdfs/PIP_Teens_Content_ Creation.pdf).

[25] D. Rushkoff, Playing the future: What we can learn from digital kids, Harper Collins, NY, 1996.

[26] Gartner Consulting, "2006 Emerging Technologies Hype Cycle", in: A. Gonsalves, TechWeb, Aug 9, 2006, (http://www.informationweek.com/internet/showArticle .jhtml?articleID=191900919).

[27] I. Ansoff, Managing strategic surprise by response to weak signals, Calif. Manage. Rev. 17 (2) (1975) 21–33.

[28] T. Könnölä, V. Brummer, A. Salo, "Diversity in foresight: Insights from the fostering of innovation ideas", Technological Forecasting & Social Change 74 (2007) 608–626.

[29] L. A. Adamic, O. Buyukkokten & E. Adar, A social network caught in the web, 2005, (http://www.cond.org/social.pdf).

[30] J, Heer & d. Boyd, Vizter: Visualizing Online Social Networks, 2005, (http://jheer.org/publications/2005-Vizster-InfoVis.pdf).

[31] European Network and Information Security Agency (ENISA) Position Paper No.1, "Security Issues and Recommendations for Online Social Networks", Editor: Giles Hogben, October 2007, http://www.enisa.europa.eu/doc/pdf/deliverables/ enisa_pp_social_networks.pdf

[32] S. Rosenbloom, "On Facebook, Scholars Link Up With Data", The New York Times, 17 December 2007, http://www.nytimes.com/2007/12/17/style/ 17facebook.html?ex=1355547600&en=3bf4c3e08da97120&ei=5088&partner=rssnyt& emc=rss

[33] Wikipedia, IBM WebFountain, http://en.wikipedia.org/wiki/WebFountain

[34] Alves J., Marques M. J., Saur I., Marques P., (2007), "Creativity and Innovation through Multi-disciplinary and Multi-sectoral Cooperation", Creativity and Innovation Management, Vol. 16, No. 1, p. 27-34.

[35] Ekvall G., (1997), "Organisational Conditions and Levels of Creativity", Creativity and Innovation Management, Vol. 6, No. 4, p. 195-205.

[36] Walton A. P., (2003), "The impact of interpersonal factors on creativity", International Journal of Entrepreneurial Behaviour & Research, Vol. 9, No. 4, p. 146-162.

[37] Wikipedia, http://en.wikipedia.org/wiki/Creativity, last accessed December 2007

[38] Fischer G., (2005), "Distances and Diversity: Sources of Social Creativity", Proceedings of the 5th conference on Creativity & Cognition, April 12-15, London, UK, p. 128-136.

[39] Shalley C. E., Zhou J., Oldham G. R., (2004), "The Effects of Personal and Contextual Characteristics on Creativity: Where Should We Go from Here?", Journal of Management, Vol. 30, No. 6, p. 933–958.

[40] BBC - British Broadcasting Corporation, "Wordsmiths hail podcast success: The term 'podcast' has been declared Word of the Year by the New Oxford American Dictionary", London 7/12/2005, (http://news.bbc.co.uk/2/hi/technology/4504256.stm).

[41] H. Bloom, Global Brain: "The Evolution of Mass Mind from the Big Bang to the 21st Century", John Wiley & Sons, Inc, New York, 2000.

[42] G. Pór, Blog of Collective Intelligence, (http://www.community- intelligence.com/blogs/public/).

[43] W. Bibel, D. Andler, O. Da Costa, G. Küppers, I.D. Pearson: "Converging Technologies and the Natural, Social and Cultural world", A report to the European

Commission from an expert group on Foresighting the new technology wave, 28 June 2004, http://europa.eu.int/comm/research/conferences/2004/ntw/pdf/sig4_en.pdf.

[44] D. Tapsott & A. D. Williams, Wikinomics: How mass collaboration changes everything, Portfolio Hardcover, New York, 2006.

[45] Benkler Y., (2006), "The Wealth of Networks. How Social Production Transforms Markets and Freedom", Yale University Press, New Haven and London, http://www.google.com/search?hl=en&q=the+wealth+of+networks (last visited August 2007)

[46] von Hippel E., (2002), "Horizontal innovation networks – by and for users", MIT Sloan School of Management Working Paper, No. 4366-02, June 2002.

[47] Wikipedia, Mind map, http://en.wikipedia.org/wiki/Mind_map

[48] T. Gordon & A. Pease, RT Delphi: An efficient "round-less" almost real time Delphi method, Technological Forecasting and Social change 73 (2006) 321-333.

[49] The Economist print edition, "The wiki principle", 20 April 2006, http://www.economist.com/surveys/displaystory.cfm?story_id=6794228

[50] Wikipedia, Digg, http://en.wikipedia.org/wiki/Digg

[51] J. C. Glenn, *14. Participatory Methods*, in: C. Glenn & T. J. Gordon (eds), Futures research methodology, AC/UNU Millennium Project, Version 2.0 2003, (http://www.acunu.org/millennium/FRM-v2.html).

[52] J. Brodkin, "IBM virtual world defies laws of physics", Network World, 20 December 2007, (http://www.networkworld.com/news/2007/122007-ibm-virtual-world.html?page=1)

[53] BBC - British Broadcasting Corporation, "Nasa investigates virtual space", 18 January 2008, (http://news.bbc.co.uk/2/hi/technology/7195718.stm)

[54] H. Hickey, "Vacation photos create 3D models of world landmarks", UW Office of News and Information, 2007, (http://uwnews.org/article.asp?articleID=37724.

In: Future of the Internet: Social Networks...   ISBN: 978-1-61209-597-4
Editors: Rick D. Sullivan and Dominick P. Bartell   ©2011 Nova Science Publishers, Inc.

Chapter 2

PERSONAL INFORMATION AGENCY AND RESELLER ADHERENCE TO KEY PRIVACY PRINCIPLES[*]

United States Government Accountability Office

WHAT GAO FOUND

In fiscal year 2005, the Departments of Justice, Homeland Security, and State and the Social Security Administration reported that they used personal information obtained from resellers for a variety of purposes. Components of the Department of Justice (the largest user of resellers) used such information in performing criminal investigations, locating witnesses and fugitives, researching assets held by individuals of interest, and detecting prescription drug fraud. The Department of Homeland Security used reseller information for immigration fraud detection and border screening programs. Uses by the Social Security Administration and the Department of State were to prevent and detect fraud, verify identity, and determine eligibility for benefits. The agencies spent approximately $30 million on contractual arrangements with resellers that enabled the acquisition and use of such information. About 91 percent of the planned fiscal year 2005 spending was for law enforcement (69 percent) or counterterrorism (22 percent).

The major information resellers that do business with the federal agencies we reviewed have practices in place to protect privacy, but these measures are not fully consistent with the Fair Information Practices. For example, the principles that the collection and use of personal information should be limited and its intended use specified are largely at odds with the nature of the information reseller business, which presupposes that personal information can be made available to multiple customers and for multiple purposes. Resellers said they believe it is not appropriate for them to fully adhere to these principles because they do not obtain their information directly from individuals. Nonetheless, in many cases, resellers take steps that address aspects of the Fair Information Practices. For example, resellers reported

---

[*] This is an edited, reformatted and augmented edition of a United States Government Accountability Office publication, Report GAO-06-42 1, dated April 2006.

that they have taken steps recently to improve their security safeguards, and they generally inform the public about key privacy principles and policies. However, resellers generally limit the extent to which individuals can gain access to personal information held about themselves, as well as the extent to which inaccurate information contained in their databases can be corrected or deleted.

Agency practices for handling personal information acquired from information resellers did not always fully reflect the Fair Information Practices. That is, some of these principles were mirrored in agency practices, but for others, agency practices were uneven. For example, although agencies issued public notices on information collections, these did not always notify the public that information resellers were among the sources to be used. This practice is not consistent with the principle that individuals should be informed about privacy policies and the collection of information. Contributing to the uneven application of the Fair Information Practices are ambiguities in guidance from the Office of Management and Budget (OMB) regarding the applicability of privacy requirements to federal agency uses of reseller information. In addition, agencies generally lack policies that specifically address these uses.

WHY GAO DID THIS STUDY

Federal agencies collect and use personal information for various purposes, both directly from individuals and from other sources, including information resellers— companies that amass and sell data from many sources. In light of concerns raised by recent security breaches involving resellers, GAO was asked to determine how the Departments of Justice, Homeland Security, and State and the Social Security Administration use personal data from these sources. In addition, GAO reviewed the extent to which information resellers' policies and practices reflect the Fair Information Practices, a set of widely accepted principles for protecting the privacy and security of personal data. GAO also examined agencies' policies and practices for handling personal data from resellers to determine whether these reflect the Fair Information Practices.

WHAT GAO RECOMMENDS

The Congress should consider the extent to which resellers should adhere to the Fair Information Practices. In addition, GAO is making recommendations to OMB and the four agencies to establish policy to address agency use of personal information from commercial sources.

Agency officials generally agreed with the content of this chapter. Resellers questioned the applicability of the Fair Information Practices, especially with regard to public records.

ABBREVIATIONS

APEC	Asia-Pacific Economic Cooperation
ATF	Bureau of Alcohol, Tobacco, Firearms, and Explosives
CBP	Customs and Border Protection
DEA	Drug Enforcement Administration
DHS	Department of Homeland Security
FBI	Federal Bureau of Investigation
FEDLINK	Federal Library and Information Network
FEMA	Federal Emergency Management Agency
FISMA	Federal Information Security Management Act
FTTTF	Foreign Terrorist Tracking Task Force
GSA	General Services Administration
ICE	Immigration and Customs Enforcement
OECD	Organization for Economic Cooperation and Development
OIG	Office of the Inspector General
OMB	Office of Management and Budget
PIA	privacy impact assessment
SSA	Social Security Administration
TSA	Transportation Security Administration
USCIS	Citizenship and Immigration Services

April 4, 2006
Congressional Committees:

Recent security breaches at large information resellers, such as ChoicePoint and LexisNexis, have highlighted the extent to which such companies collect and disseminate personal information. [1] Information resellers are companies that collect information, including personal information about consumers, from a wide variety of sources for the purpose of reselling such information to their customers, which include both private-sector businesses and government agencies. Before advanced computerized techniques made aggregating and disseminating such information relatively easy, much personal information was less accessible, being stored in paper-based public records at courthouses and other government offices or in the files of nonpublic businesses. However, information resellers have now amassed extensive amounts of personal information about large numbers of Americans, and federal agencies access this information for a variety of reasons. Federal agency use of such information is governed primarily by the Privacy Act of 1974, [2] which requires that the use of personal information be limited to predefined purposes and involve only information germane to those purposes.

The provisions of the Privacy Act are largely based on a set of principles for protecting the privacy and security of personal information, known as the Fair Information Practices, which were first proposed in 1973 by a U.S. government advisory committee [3].

These principles, now widely accepted, include:

- collection limitation,
- data quality,
- purpose specification,
- use limitation,
- security safeguards,
- openness,
- individual participation, and
- accountability. [4]

These principles, with some variation, are used by organizations to address privacy considerations in their business practices and are also the basis of privacy laws and related policies in many countries, including the United States, Germany, Sweden, Australia, New Zealand, and the European Union.

Given recent events involving information resellers and federal agencies' use of information obtained from these resellers, you asked us to review how selected federal agencies use such information. Specifically, our objectives were to determine (1) how the Departments of Justice, Homeland Security (DHS), and State and the Social Security Administration (SSA) are making use of personal information obtained through contracts with information resellers; (2) the extent to which information resellers providing personal information to these agencies have policies and practices in place that reflect the Fair Information Practices; and (3) the extent to which these agencies have policies and practices in place for the handling of personal data from resellers that reflect the Fair Information Practices.

To address our first objective, we analyzed fiscal year 2005 contracts and other vehicles for the acquisition of personal information from information resellers by DHS, Justice, State, and SSA to identify their purpose, scope, and value. We obtained additional information on these contracts and uses in discussions with agency officials to ensure that all relevant information had been provided to us.

To address our second objective, we reviewed documentation from five major information resellers [5] and conducted site visits at three of them [6] to obtain information on privacy and security policies and procedures and compared these with the Fair Information Practices. In conducting our analysis, we identified the extent to which reseller practices were consistent with the key privacy principles of the Fair Information Practices. We also assessed the potential effect of any inconsistencies; however, we did not attempt to make determinations of whether or how information reseller practices should change. Such determinations are a matter of policy based on balancing the public's right to privacy with the value of services provided by resellers to customers such as government agencies. We determined that the five resellers we reviewed accounted for most of the contract value of personal information obtained from resellers in fiscal year 2005 by the four agencies we reviewed. We did not evaluate the effectiveness of resellers' information security programs.

To address our third objective, we identified and evaluated agency guidelines and management policies and procedures governing the use of personal information obtained from information resellers and compared these to the Fair Information Practices. We also conducted interviews at the four agencies with senior agency officials designated for privacy issues as well as officials of the Office of Management and Budget (OMB) to obtain their views on the applicability of federal privacy laws and related guidance to agency use of information resellers. We performed our work from May 2005 to March 2006 in the Washington, D.C., metropolitan area; Little Rock, Arkansas; Alpharetta, Georgia; and Miamisburg, Ohio. Our work was performed in accordance with generally accepted government auditing standards. Our objectives, scope, and methodology are discussed in more detail in appendix I.

RESULTS IN BRIEF

In fiscal year 2005, Justice, DHS, State, and SSA reported using personal information from information resellers for a variety of purposes, including law enforcement, counterterrorism, fraud prevention, and debt collection. Taken together, approximately 91 percent of planned spending on resellers reported by the agencies for fiscal year 2005 was for law enforcement (69 percent) or counterterrorism (22 percent). For example, components of the Department of Justice (the largest user of resellers) made use of such information for criminal investigations, location of witnesses and fugitives, research of assets held by individuals of interest, and detection of fraud in prescription drug transactions. Examples of uses by the DHS include immigration fraud detection and border screening programs. SSA and State acquire personal information from information resellers for fraud detection and investigation, identity verification, and benefit eligibility determination. The four agencies obtained personal information from resellers primarily through two general-purpose governmentwide contract vehicles—the Federal Supply Schedule of the General Services Administration (GSA) and the Library of Congress's Federal Library and Information Network. Collectively, the four agencies reported approximately $30 million [7] in fiscal year 2005 in contractual arrangements with information resellers that enabled the acquisition and use of personal information.

The major information resellers that do business with the federal agencies we reviewed have practices in place to protect privacy, but these measures are not fully consistent with the Fair Information Practices. For example, the nature of the information reseller business is largely at odds with the principles of collection limitation, data quality, purpose specification, and use limitation. These principles center on limiting the collection and use of personal information, and they link data quality (e.g., accuracy) requirements to these limitations. Resellers said they believe it may not be appropriate or practical for them to fully adhere to these principles because they do not obtain their information directly from individuals. In fact, the information reseller industry is based on multipurpose collection and use of personal and other information [8] information from multiple sources. In many cases, resellers take steps that address aspects of the Fair Information Practices. For example, resellers reported that they have taken steps recently to improve their security safeguards, and they generally inform the public about key privacy principles and policies (relevant to the openness principle). However, resellers generally limit the extent to which individuals can gain access to personal information held about themselves as well as the extent to which inaccurate information contained in their databases can be corrected or deleted (relevant to the individual participation principle).

Agency practices for handling personal information acquired from information resellers reflected the principles of the Fair Information Practices in four cases and in the other four did not. Specifically, regarding the collection limitation, data quality, use limitation, and security safeguards principles, agency practices generally reflected the Fair Information Practices. For example, regarding the data quality principle that data should be accurate, current, and complete, as needed for the defined purpose, law enforcement agencies (including the Federal Bureau of Investigation and the U.S. Secret Service) generally reported that they corroborate information obtained from resellers to ensure that it is accurate when it is used as part of an investigation.

Regarding other principles, however, agency practices were uneven. Specifically, agencies did not always have practices in place to fully address the purpose specification, individual participation, openness, and accountability principles with regard to use of reseller information. For example,

- although agencies notify the public through Federal Register notices and published privacy impact assessments that they collect personal information from various sources, they do not always indicate specifically that information resellers are among those sources, and
- some agencies lack robust audit mechanisms to ensure that use of personal information from information resellers is for permissible purposes, reflecting an uneven application of the accountability principle.

Contributing to the uneven application of the Fair Information Practices are ambiguities in guidance from OMB regarding the applicability of privacy requirements to federal agency uses of reseller information. In addition, agencies generally lack policies that specifically address these uses.

The Congress should consider the extent to which information resellers should adhere to the Fair Information Practices. We are also recommending that the Director, OMB, revise privacy guidance to clarify the applicability of requirements for public notices and privacy impact assessments to agency use of personal information from resellers and direct agencies to review their uses of such information to ensure it is explicitly referenced in privacy notices and assessments. Further, we are recommending that agencies develop specific policies for the use of personal information from resellers.

We obtained written comments on a draft of this chapter from Justice, DHS, SSA, and State. We also received comments via E-mail from OMB. Comments from Justice, DHS, SSA, and State are reproduced in appendixes III to VI, respectively. Justice, DHS, SSA, and OMB all generally agreed with the chapter and described actions initiated to address our recommendations. In its comments, Justice recommended that prior to issuance of any new or revised policy, careful consideration be given to its impact on Justice. We believe the policy clarifications we are proposing are unlikely to result in an adverse impact on law enforcement activities at Justice. Justice and SSA also provided technical comments, which were incorporated in the final chapter as appropriate.

State interpreted our draft chapter to "rest on the premise that records from 'information resellers' should be accorded special treatment when compared with sensitive information from other sources." State also indicated that it does not distinguish between types of information or sources of information in complying with privacy laws. However, our chapter does not suggest that data from resellers should receive special treatment. Instead, our chapter takes the widely accepted Fair Information Practices as a universal benchmark of privacy protections and assesses agency practices in comparison with them.

We also obtained comments on excerpts of our draft chapter from the five information resellers we reviewed. Several resellers raised concerns regarding the version of the Fair Information Practices we used to assess their practices, stating their view that it was more appropriate for organizations that collection information directly from consumers and that they were not legally bound to adhere to the Fair Information Practices. As discussed in our chapter, the version of the Fair Information Practices we used has been widely adopted and cited within the federal government as well as internationally. Further, we use it as an analytical framework for identifying potential privacy issues for further consideration by Congress— not as criteria for strict compliance. Resellers also stated that the draft did not take into account that public record information is open to all for any use not prohibited by state or federal law. However, we believe it is not clear that individuals give up all privacy rights to personal information contained in public records, and we believe it is important to assess the status of privacy protections for all personal information being offered commercially to the government so that informed policy decision can be made about the appropriate balance between resellers' services and the public's right to privacy. Resellers also offered technical comments, which were incorporated in the final chapter as appropriate.

BACKGROUND

Before advanced computerized techniques for aggregating, analyzing, and disseminating data came into widespread use, personal information contained in paper-based public records at courthouses or other government offices was relatively difficult to obtain, usually requiring a personal visit to inspect the records. Nonpublic information, such as personal information contained in product registrations, insurance applications, and other business records, was also generally inaccessible. In recent years, however, advances in technology have spawned information reseller businesses that systematically collect extensive amounts of personal information from a wide variety of sources and make it available electronically over the Internet and by other means to customers in both government and the private sector. This automation of the collection and aggregation of multiple-source data, combined with the ease and speed of its retrieval, have dramatically reduced the time and effort needed to obtain information of this type. Among the primary customers of information resellers are financial institutions (including insurance companies), retailers, law offices, telecommunications and technology companies, and marketing firms.

We use the term "information resellers" to refer to businesses that vary in many ways but have in common the fact that they collect and aggregate personal information from multiple sources and make it available to their customers. These businesses do not all focus exclusively on aggregating and reselling personal information. For example, Dun and Bradstreet primarily provides information on commercial enterprises for the purpose of contributing to decision making regarding those enterprises. In doing so, it may supply personal information about individuals associated with those commercial enterprises. To a certain extent, the activities of information resellers may also overlap with the functions of consumer reporting agencies, also known as credit bureaus—entities that collect and sell information about individuals' creditworthiness, among other things. As is discussed further below, to the extent that information resellers perform the functions of consumer reporting agencies, they are subject to legislation specifically addressing that industry, particularly the Fair Credit Reporting Act.

Information resellers obtain personal information from many different sources. Generally, three types of information are collected: public records, publicly available information, and nonpublic information.

- Public records are a primary source of information about consumers, available to anyone, and can be obtained from governmental entities. What constitutes public records is dependent upon state and federal laws, but generally these include birth and death records, property records, tax lien records, motor vehicle registrations, voter registrations, licensing records, and court records (including criminal records, bankruptcy filings, civil case files, and legal judgments).
- Publicly available information is information not found in public records but nevertheless publicly available through other sources. These sources include telephone directories, business directories, print publications such as classified ads or magazines, Internet sites, and other sources accessible by the general public.
- Nonpublic information is derived from proprietary or nonpublic sources, such as credit header data, [9] product warranty registrations, and other application information provided to private businesses directly by consumers.

Private sector businesses rely on information resellers for information to support a variety of activities, such as

- conducting pre-employment background checks on prospective employees,
- verifying individuals' identities by reviewing records of their personal information;
- marketing commercial products to consumers matching specified demographic characteristics; and
- preventing financial fraud by examining insurance, asset, and other financial record information.

Typically, while information resellers may collect and maintain personal information in a variety of databases, they provide their customers with a single, consolidated online source for a broad array of personal information. Figure 1 illustrates how information is collected from multiple sources and ultimately accessed by customers, including government agencies, through contractual agreements.

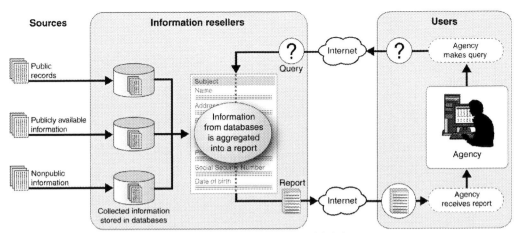

Source: GAO analysis of information reseller and agency-provided data.

Figure 1. Typical Information Flow through Resellers to Government Customers.

In addition to providing consolidated access to personal information through Internet-based Web sites, information resellers offer a variety of products tailored to the specific needs of various lines of business. For example, an insurance company could obtain different products covering police and accident reports, insurance carrier information, vehicle owner verification or claims history, or online public records. Typically, services offered to law enforcement officers include more information—including sensitive information, such as full Social Security numbers and driver's license numbers—than is offered to other customers.

Federal Laws and Guidance Govern Use of Personal Information in Federal Agencies

There is no single federal law that governs all use or disclosure of personal information. Instead, U.S. law includes a number of separate statutes that provide privacy protections for information used for specific purposes or maintained by specific types of entities. The major requirements for the protection of personal privacy by federal agencies come from two laws, the Privacy Act of 1974 and the privacy provisions of the E-Government Act of 2002. The Federal Information Security Management Act of 2002 (FISMA) also addresses the protection of personal information in the context of securing federal agency information and information systems.

The Privacy Act places limitations on agencies' collection, disclosure, and use of personal information maintained in systems of records. The act describes a "record" as any item, collection, or grouping of information about an individual that is maintained by an agency and contains his or her name or another personal identifier. It also defines "system of records" as a group of records under the control of any agency from which information is retrieved by the name of the individual or by an individual identifier. The Privacy Act requires that when agencies establish or make changes to a system of records, they must notify the public by a notice in the Federal Register identifying, among other things, the type of data collected, the types of individuals about whom information is collected, the intended "routine" uses of data, and procedures that individuals can use to review and correct personal information. [10]

The act's requirements also apply to government contractors when agencies contract for the development and maintenance of a system of records to accomplish an agency function. [11] The act limits its applicability to cases in which systems of records are maintained specifically on behalf of a government agency.

Several provisions of the act require agencies to define and limit themselves to specific predefined purposes. For example, the act requires that to the greatest extent practicable, personal information should be collected directly from the subject individual when it may affect an individual's rights or benefits under a federal program. The act also requires that an agency inform individuals whom it asks to supply information of (1) the authority for soliciting the information and whether disclosure of such information is mandatory or voluntary; (2) the principal purposes for which the information is intended to be used; (3) the routine uses that may be made of the information; and (4) the effects on the individual, if any, of not providing the information. According to OMB, this requirement is based on the assumption that individuals should be provided with sufficient information about the request to make a decision about whether to respond.

In handling collected information, the Privacy Act also requires agencies to, among other things, allow individuals to (1) review their records (meaning any information pertaining to them that is contained in the system of records), (2) request a copy of their record or information from the system of records, and (3) request corrections in their information. Such provisions can provide a strong incentive for agencies to correct any identified errors.

Agencies are allowed to claim exemptions from some of the provisions of the Privacy Act if the records are used for certain purposes. For example, records compiled for criminal law enforcement purposes can be exempt from a number of provisions, including (1) the requirement to notify individuals of the purposes and uses of the information at the time of collection and (2) the requirement to ensure the accuracy, relevance, timeliness, and completeness of records. A broader category of investigative records compiled for criminal or civil law enforcement purposes can also be exempted from a somewhat smaller number of Privacy Act provisions, including the requirement to provide individuals with access to their records and to inform the public of the categories of sources of records. In general, the exemptions for law enforcement purposes are intended to prevent the disclosure of information collected as part of an ongoing investigation that could impair the investigation or allow those under investigation to change their behavior or take other actions to escape prosecution.

The E-Government Act of 2002 strives to enhance protection for personal information in government information systems or information collections by requiring that agencies conduct privacy impact assessments (PIA). A PIA is an analysis of how personal information is collected, stored, shared, and managed in a federal system. More specifically, according to OMB guidance, [12] a PIA is an analysis of how

> ...information is handled: (i) to ensure handling conforms to applicable legal, regulatory, and policy requirements regarding privacy; (ii) to determine the risks and effects of collecting, maintaining, and disseminating information in identifiable form in an electronic information system; and (iii) to examine and evaluate protections and alternative processes for handling information to mitigate potential privacy risks.

Agencies must conduct PIAs (1) before developing or procuring information technology that collects, maintains, or disseminates information that is in a personally identifiable form or (2) before initiating any new data collections involving personal information that will be collected, maintained, or disseminated using information technology if the same questions are asked of 10 or more people. OMB guidance also requires agencies to conduct PIAs when a system change creates new privacy risks, for example, changing the way in which personal information is being used. The requirement does not apply to all systems. For example, no assessment is required when the information collected relates to internal government operations, the information has been previously assessed under an evaluation similar to a PIA, or when privacy issues are unchanged.

FISMA also addresses the protection of personal information. FISMA defines federal requirements for securing information and information systems that support federal agency operations and assets; it requires agencies to develop agencywide information security programs that extend to contractors and other providers of federal data and systems. [13] Under FISMA, information security means protecting information and information systems from unauthorized access, use, disclosure, disruption, modification, or destruction, including controls necessary to preserve authorized restrictions on access and disclosure to protect personal privacy, among other things.

OMB is tasked with providing guidance to agencies on how to implement the provisions of the Privacy Act and the E-Government Act and has done so, beginning with guidance on the Privacy Act, issued in 1975. [14] The guidance provides explanations for the various provisions of the law as well as detailed instructions for how to comply. OMB's guidance on implementing the privacy provisions of the E-Government Act of 2002 identifies circumstances under which agencies must conduct PIAs and explains how to conduct them. OMB has also issued guidance on implementing the provisions of FISMA.

Additional Laws Provide Privacy Protections for Specific Types and Uses of Information

Although federal laws do not specifically regulate the information reseller industry as a whole, they provide safeguards for personal information under certain specific circumstances, such as when financial or health information is involved, or for such activities as pre-employment background checks.

Table 1. Federal Laws Addressing Private Sector Disclosure of Personal Information

Federal laws	Provisions
Fair Credit Reporting Act	Consumer reporting agencies are limited to providing data onlyto their customers that have a permissible purpose for using thedata. With few exceptions, government agencies are treated like other parties and must have a permissible purpose in orderto obtain a consumer report.

Table 1 (Continued)

Federal laws	Provisions
Gramm-Leach-Bliley Act	Sets limitationson financial institutions' disclosure of customer data to third parties, such as information resellers. Requires companies to give consumers privacy notices that explain the institutions' information-sharing practices. In turn, consumers have the right to limit some, but not all, sharing of their nonpublic personal information.
Driver's Privacy Protection Act	Restricts a third party's ability to obtain Social Security numbersand other driver's license information from state motor vehicle offices unless doing so for a permissible purpose under the law;restricts state motor vehicle offices' ability to disclose driver's license information.
Health Insurance Portability and Accountability Act	Health care organizations are restricted from disclosing a patient's health information without the patient's consent, exceptfor permissible reasons, and are required to inform individuals of privacy practices.
Fair and Accurate Credit Transactions Act	Consumers may obtain one free annual consumer report from nationwide consumer reporting agencies.

Source: GAO analysis.
Note: Appendix II provides additional details on the requirements of these laws.

Specifically, the Fair Credit Reporting Act, the GrammLeach-Bliley Act, the Driver's Privacy Protection Act, and the Health Insurance Portability and Accountability Act all restrict the ways in which businesses, including information resellers, may use and disclose consumers' personal information (see app. II for more details about these laws). The Gramm-Leach-Bliley Act, for example, limits financial institutions' disclosure of nonpublic personal information to nonaffiliated third parties and requires companies to give consumers privacy notices that explain the institutions' information sharing practices. Consumers then have the right to limit some, but not all, sharing of their nonpublic personal information.

As shown in table 1, these laws either restrict the circumstances under which entities such as information resellers are allowed to disclose personal information or restrict the parties with whom they are allowed to share information.

Information resellers are also affected by various state laws. For example, California state law requires businesses to notify consumers about security breaches that could directly affect them. Legal requirements, such as the California law, led ChoicePoint, a large information reseller, to notify its customers in mid-February 2005 of a security breach in which unauthorized persons gained access to personal information from its databases. Since the ChoicePoint notification, bills were introduced in at least 35 states and enacted in at least 22 states [15] that require some form of notification upon a security breach.

The Fair Information Practices Are Widely Agreed to Be Key Principles for Privacy Protection

The Fair Information Practices are a set of internationally recognized privacy protection principles. First proposed in 1973 by a U.S. government advisory committee, the Fair Information Practices were intended to address what the committee termed a poor level of protection afforded to privacy under contemporary law. [16] A revised version of the Fair Information Practices, developed by the Organization for Economic Cooperation and Development (OECD) [17] in 1980, has been widely adopted. The OECD principles are shown in table 2.

The Fair Information Practices are, with some variation, the basis of privacy laws and related policies in many countries, including the United States, Germany, Sweden, Australia, New Zealand, and the European Union. [18] They are also reflected in a variety of federal agency policy statements, beginning with an endorsement of the OECD principles by the Department of Commerce in 1981, [19] and including policy statements of the DHS, Justice, Housing and Urban Development, and Health and Human Services. [20] In 2004, the Chief Information Officers Council issued a coordinating draft of their Security and Privacy Profile for the Federal Enterprise Architecture [21] that links privacy protection with a set of acceptable privacy principles corresponding to the OECD's version of the Fair Information Practices.

The Fair Information Practices are not precise legal requirements. Rather, they provide a framework of principles for balancing the need for privacy with other public policy interests, such as national security, law enforcement, and administrative efficiency. Striking that balance varies among countries and among types of information (e.g., medication versus employment information).

Table 2. The OECD Fair Information Practices

Principle	Description
Collection limitation	The collection of personal information should be limited, should be obtained by lawful and fair means, and, where appropriate, with the knowledge or consent of the individual.
Data quality	Personal information should be relevant to the purpose for which it is collected, and should be accurate, complete, and current as needed for that purpose.
Purpose specification	The purposes for the collection of personal information should be disclosed before collection and upon any change to that purpose, and its use should be limited to those purposes and compatible purposes.
Use limitation	Personal information should not be disclosed or otherwise used for other than a specified purpose without consent of the individual or legal authority.
Security safeguards	Personal information should be protected with reasonable security safeguards against risks such as loss or unauthorized access, destruction, use, modification, or disclosure.
Openness	The public should be informed about privacy policies and practices, and individuals should have ready means of learning about the use of personal information.

Table 2 (Continued)

Principle	Description
Individual participation	Individuals should have the following rights: to know about the collection of personal information, to access that information, to request correction, and to challenge the denial of those rights.
Accountability	Individuals controlling the collection or use of personal information should be accountable for taking steps to ensure theimplementation of these principles.

Source: OECD.

The Fair Information Practices also underlie the provisions of the Privacy Act of 1974. For example, the system of records notice required under the Privacy Act embodies the purpose specification, openness, and individual participation principles in that it provides a public accounting through the Federal Register of the purpose and uses for personal information, and procedures by which individuals may access and correct, if necessary, information about themselves. Further, the E-Government Act's requirement to conduct PIAs likewise reflects the Fair Information Practices. Under the act, agencies are to make these assessments publicly available, if practicable, through agency Web sites or by publication in the Federal Register, or other means. To the extent that such assessments are made publicly available, they also provide notice to the public about the purpose of planned information collections and the planned uses of the information being collected.

Congressional Interest in the Information Reseller Industry Has Been Heightened

A number of congressional hearings were held and bills introduced in 2005 in the wake of widely publicized data security breaches at major information resellers such as ChoicePoint and LexisNexis as well as other firms. In March 2005, the House Subcommittee on Commerce, Trade, and Consumer Protection of the House Energy and Commerce Committee held a hearing entitled "Protecting Consumers' Data: Policy Issues Raised by ChoicePoint," which focused on potential remedies for security and privacy concerns regarding information resellers. Similar hearings were held by the House Energy and Commerce Committee and by the U.S. Senate Committee on Commerce, Science, and Transportation in spring 2005.

The heightened interest in this subject led a number of Members of Congress to propose a variety of bills aimed at regulating companies that handle personal information, including information resellers. Several of these bills require companies such as information resellers to notify the public of security breaches, while a few also allow consumers to "freeze" their credit (i.e., prevent new credit accounts from being opened without special forms of authentication), or see and correct personal information contained in reseller data collections. Other proposed legislation includes (1) the Data Accountability and Trust Act, [22] requiring security policies and procedures to protect computerized data containing personal information and nationwide notice in the event of a security breach, and (2) the Personal Data Privacy and

Security Act of 2005, [23] requiring data brokers to disclose personal electronic records pertaining to an individual and inform individuals on procedures for correcting inaccuracies.

USING GOVERNMENTWIDE CONTRACTS, FEDERAL AGENCIES OBTAIN

Personal Information from Information Resellers for a Variety of Purposes

Primarily through governmentwide contracts, Justice, DHS, State, and SSA reported using personal information obtained from resellers for a variety of purposes, including law enforcement, counterterrorism, fraud detection/prevention, and debt collection. Most uses by Justice were for law enforcement and counterterrorism, such as investigations of fugitives and obtaining information on witnesses and assets held by individuals of interest. DHS also used reseller information primarily for law enforcement and counterterrorism, such as screening vehicles entering the United States. State and SSA reported acquiring personal information from information resellers for fraud detection and investigation, identity verification, and benefit eligibility determination. The four agencies reported approximately $30 million in contractual arrangements with information resellers in fiscal year 2005. [24] Justice accounted for most of the funding (about 63 percent). Approximately 91 percent of agency uses of reseller data were in the categories of law enforcement (69 percent) or counterterrorism (22 percent). Figure 2 details contract values categorized by their reported use. (Details on uses by each agency are given in the individual agency discussions.)

Department of Justice Uses Information Resellers Primarily for Law Enforcement and Counterterrorism Purposes

According to Justice contract documentation, access to up-to-date and comprehensive public record information is a critical ongoing mission requirement, and the department relies on a wide variety of information resellers—including ChoicePoint, Dun and Bradstreet, LexisNexis, and West—to meet that need. Departmental use of information resellers was primarily for purposes related to law enforcement (75 percent) and counterterrorism (18 percent), including support for criminal investigations, location of witnesses and fugitives, information on assets held by individuals under investigation, and detection of fraud in prescription drug transactions. In fiscal year 2005, Justice and its components reported approximately $19 million in acquisitions from information resellers involving personal information. The department acquired these services primarily through use of GSA's Federal SupplySchedule [25] offerings including a blanket purchase agreement [26] with ChoicePoint valued at approximately $15 million. [27] Several component agencies, such as the Federal Bureau of Investigation (FBI), the Drug Enforcement Administration (DEA), and the Bureau of Alcohol, Tobacco, Firearms, and Explosives (ATF) placed orders with information resellers based on the schedules. In addition, for fiscal year 2005, Justice established separate departmentwide contracts with LexisNexis and West valued at $4.5 million and $5.2 million, respectively. [28]

Tasked to protect and defend the United States against terrorist and foreign intelligence threats and to enforce criminal laws, the FBI is Justice's largest user of information resellers, with about $11 million in contracts in fiscal year 2005. The majority of FBI's use involves two major programs, the Public Source Information Program and the Foreign Terrorist Tracking Task Force (FTTTF).

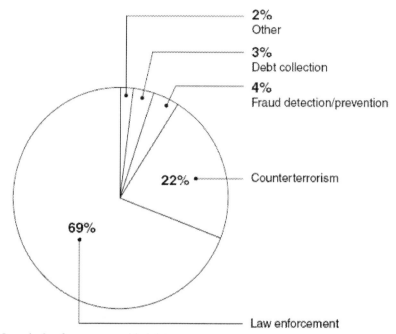

Source: GAO analysis of agency-provided data.

Figure 2. Fiscal Year 2005 Contractual Vehicles Enabling the Use of Personal Information from Information Resellers, Categorized by Reported Use.

In support of the investigative and intelligence missions of the FBI, the Public Source Information Program provides all offices of the FBI with access via the Internet to public record, legal, and news media information available from various online commercial databases. These databases are used to assist with investigations by identifying the location of individuals and identifying alias names, Social Security numbers, relatives, dates of birth, telephone numbers, vehicles, business affiliations, other associations, and assets. Public Source Information Program officials reported that use of these commercial databases often results in new information regarding the subject of the investigation. Officials noted that commercial databases are used in preliminary investigations, and that subsequently, investigative personnel must verify the results of each search.

The FBI's FTTTF also contracts with several information resellers (1) to assist in fulfilling its mission of assisting federal law enforcement and intelligence agencies in locating foreign terrorists and their supporters who are in or have visited the United States and (2) to provide information to other law enforcement and intelligence community agencies that can lead to their surveillance, prosecution, or removal. As we previously reported, [29] FTTTF makes use of personal information from several commercial sources to analyze intelligence and detect terrorist activities in support of ongoing investigations by law enforcement agencies and the intelligence community. Information resellers provide FTTTF with names, addresses, telephone numbers, and other biographical and demographical information as well as legal briefs, vehicle and boat registrations, and business ownership records.

Other Justice components reported using personal information from information resellers to support the conduct of investigations and other law enforcement-related activities. For example, the U.S. Marshals Service uses an information reseller to, among other things, locate fugitives by identifying a fugitive's relatives and their addresses. [30] Through interviews with relatives, a U.S. Marshal may be able to ascertain the location of a fugitive and subsequently apprehend the individual.

DEA, the second largest Justice user of information resellers in fiscal year 2005, obtains reseller data to detect fraud in prescription drug transactions. [31] Through these data, DEA agents can detect irregular prescription patterns for specific drugs and trace this information to the pharmacy and prescribing doctor. [32] DEA also uses an information reseller to locate individuals in asset forfeiture cases. [33] Reseller data allows DEA to identify all possible addresses for an individual in order to meet the agency's obligation to make a reasonable effort to notify individuals of seized property and inform them of their rights to contest the seizures.

Other uses reported by Justice components are not related to law enforcement. For example, uses by the U.S. Trustees, Antitrust, Civil, Tax, and Criminal Divisions include ascertaining the financial status of individuals for debt collection purposes or bankruptcy proceedings or for the location of individuals for court proceedings. The Executive Office for U.S. Attorneys uses information resellers to ascertain the financial status of those indebted to the United States in order to assess the debtor's ability to repay the debt. According to officials, information reseller databases may reveal assets that a debtor is attempting to conceal. Further, the U.S. Attorneys use information resellers to locate victims of federal crime in order to notify these individuals of relevant court proceedings pursuant to the Justice for All Act. [34]

Table 3 details in aggregate the vendors, fiscal year 2005 contract values, and reported uses for contracts with information resellers by major Justice components.

Table 3. Reported Uses of Personal Information: Department of Justice Contracts with Information Resellers, Fiscal Year 2005

Major component	Information resellers	Aggregate contract value	Uses involving personal information
Federal Bureau of Investigation	ChoicePoint, LexisNexis, West, Credit Bureau Reports, Dun and Bradstreet, Seisinta	$11,248,000	Public Source Information Program. Find individuals and identify alias names, Social Security numbers, relatives, dates of birth, telephone numbers, vehicles, business affiliations, associations, and assets.The program provides FBI units with access to public record, legal, and news media information from various online commercial databases. Criminal Investigative Division.Same use.Foreign Terrorist Tracking Task Force. Obtain such information as names, addresses, telephone numbers, other biographical information, vehicle and boat registrations, and business ownership records.

Table 3 (Continued)

Major component	Information resellers	Aggregate contract value	Uses involving personal information
Drug Enforcement Administration	ChoicePoint, LexisNexis, Dun and Bradstreet	$4,283,000	Conduct investigations of drug diversions and improper drug transactions: For example, identifying cases in which physicians sell prescriptions to drug dealers or abusers, pharmacists falsely report legitimate drug sales and subsequently sell the drugs illegally, and employees steal from inventory and falsify orders to hide illicit sales. Support criminal investigations of specific individuals and companies.Locate an individual's address in asset removal cases.
U.S. Marshals Service	ChoicePoint, LexisNexis, West	$1,661,000	Generate leads related to fugitive investigations (e.g., a fugitive's relatives and their contact information). Asset Forfeiture Office. Obtain information on preseized, seized, and forfeited property. The Marshals Service offers property for sale to the public that has been forfeited under laws enforced or administered by Justice and its investigative agencies. Office of General Counsel. Research assets to administer tort claims against the service. For example, if a claimant makes an assertion that the service is responsible for damaging property and does not provide supporting documentation, General Counsel personnel may use commercial data to verify tax assessment records, proof of ownership, etc.
Executive Office for U.S. Attorneys	ChoicePoint, CBR Information Services	$855,000	Financial Litigation Units. Ascertain the financial status of individuals and uncover concealed assets for civil and criminal debt collection efforts. Locate and notify crime victims of relevant court proceedings pursuant to theJustice for All Act of 2004.
Bureau of Alcohol, Tobacco, Firearms, and Explosives	ChoicePoint, Dun and Bradstreet, LexisNexis, West	$791,000	Support investigative activities such as locating and apprehending fugitives or obtaining data on businesses (such as in arson investigations), which may include personal information about business owners.
Executive Office of the United States Trustees	ChoicePoint, Equifax,bReal Data Corp, MLS Hawaii	$303,000	Obtain information on assets (openly held or concealed) of individuals in bankruptcy proceedings (as part of office's mission to enforce bankruptcy laws and provide oversight of private trustees).Obtain credit reports on employees as part of a security clearance process.
Office of the Inspector General	ChoicePoint, LexisNexis, West	$43,000	Investigations Division. Support investigations of alleged violations of fraud,abuse, and integrity laws that govern Justice employees, operations, grantees, and contractors.

Major component	Information resellers	Aggregate contract value	Uses involving personal information
U.S. National Central Bureau	ChoicePoint	$31,000	Conduct business and address checks on individuals who may be potentially involved in fraud or fugitive cases.The bureau facilitates international law enforcement cooperation as the U.S. representative of the International Criminal Police Organization (INTERPOL).
National Drug Intelligence Center	ChoicePoint	$28,000	Document Exploitation Division. Locate individuals, identify assets, and investigate fraud. The Document Exploitation Division specializes in analyzing information seized in major federal drug investigations.
Office of Justice Programs	Dun and Bradstreet	$22,000	Office of Comptroller, Financial Management Division. Obtain credit reportsto assess new grantees' (nongovernmental or nontribal) financial integrity. These credit reports may include personal information on company owners.This information is used to support the new grantee's ability to operate thegrant programs of the Office of Justice Programs, to confirm the existenceof the company, and to determine any outstanding liens or obligations thatmight influence the success of the grant program.
Litigating Divisions (Civil, Criminal, Antitrust, and Tax)	ChoicePoint, Credit Bureau Reports (division of CBC Companies)	$21,000	Civil Division. Locate individuals and assets in connection with litigation for purposes such as obtaining depositions, debt collection, and identifying assets that a debtor may be concealing in bankruptcy proceedings.Criminal Division, Office of Special Investigations. Locate individuals who may have taken part in Nazi-sponsored acts of persecution abroad before and during World War II and who subsequently entered, or seek to enter, theUnited States illegally and/or fraudulently.Antitrust Division. Locate witnesses for trials.Tax Division. Obtain credit bureau reports for debt collection purposes.

Source: Department of Justice.

Notes: The table represents fiscal year 2005 contract values and may not reflect actual expenditures. We did not verify the accuracy or completeness of the dollar figures provided to us.

Contract values were rounded to the nearest thousand. Several Justice components use departmentwide contracts with LexisNexis and West, which provide, among other things, access to public records information. Several components, including the litigating divisions (Civil, Criminal, Antitrust, and Tax), the Office of Justice Programs, and the Executive Office for U.S. Attorneys, reported that their use of these departmentwide contracts was primarily for legal research, and therefore we did not include these uses in the table.

[a]Seisint is now owned by LexisNexis.

[b]Equifax is an example of a consumer reporting agency. Consumer reporting agencies, also known as credit bureaus, are entities that collect and sell information about the creditworthiness, among other things, of individuals and are required by the Fair Credit Reporting Act to disclose such information only for permissible purposes.

DHS Uses Information Resellers Primarily for Law Enforcement and Counterterrorism

In fiscal year 2005, DHS and its components reported that they used information reseller data primarily for law enforcement purposes, such as for developing leads on subjects in criminal investigations and detecting fraud in immigration benefit applications (part of enforcing the immigration laws). Counterterrorism uses involved screening programs at the northern and southern borders as well as at the nation's airports. DHS reported planning to spend about $9 million acquiring personal information from resellers in fiscal year 2005. DHS acquired these services primarily for law enforcement (63 percent) and counterterrorism (35 percent) purposes through FEDLINK—a governmentwide contract vehicle provided by the Library of Congress—and GSA's Federal Supply Schedule contracts as well as direct purchases by its components. DHS's primary vehicle for acquiring data from information resellers was the FEDLINK contract vehicle, which DHS used to acquire reseller services from Choicepoint ($4.1 million), Dun and Bradstreet ($640,000), LexisNexis ($2 million), and West ($1 million).

U.S. Immigration and Customs Enforcement (ICE) is DHS's largest user of personal information from resellers, with acquisitions worth over $4.3 million. The largest investigative component of DHS, ICE has as its mission to prevent acts of terrorism by targeting the people, money, and materials that support terrorist and criminal activities. ICE uses information resellers to collect personal information for criminal investigative purposes and to perform background security checks. Data commonly obtained include address and vehicle information; according to officials, this information is either used to verify data already collected or is itself verified by investigators through other means. For example, ICE's Federal Protective Service has about 50 users who access an information reseller database to assist in properly identifying and locating potential criminal suspects. Investigators may verify an address obtained from the database by confirming billing information with a utility company or by conducting "drive-by" surveillance. The Federal Protective Service views information obtained from resellers as "raw" or "unverified" data, which may or may not be of use to investigators.

Other DHS components likewise reported using personal information from resellers to support investigations and other law enforcement-related activities. For example, U.S. Customs and Border Protection (CBP)— tasked with managing, controlling, and protecting the nation's borders at and between the official ports of entry—uses information resellers for law enforcement, intelligence gathering, and prosecution support. Using these databases, investigators conduct queries on people, businesses, property, and corresponding links via a secure Internet connection. According to officials, information obtained is corroborated with other previously obtained data, open-source information, and investigative leads.

CBP also uses a specially developed information reseller product to assist law enforcement officials in vehicle identification at northern and southern land borders. CBP uses electronic readers to capture license plate data on vehicles entering or exiting U.S. borders, converts the data to an electronic format, and transmits the data to an information reseller, which returns U.S. motor vehicle registration information to CBP. The license plate data, merged with the associated motor vehicle registration data provided by the reseller, are then checked against government databases in order to help assess risk related to vehicles (i.e., a vehicle whose license plate is associated with a law enforcement record might be referred for secondary examination).

The Federal Emergency Management Agency (FEMA), charged with building and supporting the nation's emergency management system, uses an information reseller to detect fraud in disaster assistance applications. FEMA uses this service to verify information that individuals present in their applications for disaster assistance via the Internet. At the time of application, an individual is required to pass an identity check that determines whether the presented identity exists, followed by an identity validation quiz to better ensure that the applicant corresponds to the identity presented. The information reseller is used to verify the applicant's name, address, and Social Security number.

DHS is also using information resellers in its counterterrorism efforts. For example, the Transportation Security Administration (TSA), tasked with protecting the nation's transportation systems, used data obtained from information resellers as part of a test associated with the development of its domestic passenger prescreening program, called "Secure Flight." [35] TSA's plans for Secure Flight involve the submission of passenger information by an aircraft operator to TSA whenever a reservation is made for a flight in which the origin and destination are domestic airports. In the prescreening of airline passengers, this information would be compared with federal watch lists of individuals known or suspected of activities related to terrorism. TSA conducted a test designed to help determine the extent to which information resellers could be used to authenticate passenger identity information provided by air carriers. It plans to use the test results to determine whether commercial data can be used to improve the effectiveness of watch-list matching by identifying passengers who would not have been identified from passenger name records and government data alone. The test results also may be used to identify items of personally identifying information that should be required of passengers to improve aviation security.

Table 4 provides detailed information about DHS uses of information resellers in fiscal year 2005, as reported by officials of the department's components.

Table 4. Reported Uses of Personal Information: DHS Contracts with Information Resellers, Fiscal Year 2005

Major component	Information reseller	Aggregate contract value	Uses involving personal information
U.S. Immigration and Customs Enforcement	ChoicePoint, Dun and Bradstreet, LexisNexis, West	$4,389,00 0	Acquire data (generally, address and vehicle information) for criminal investigations and background security checks.According to officials, information is either used to verify data already collected or is itself verified by investigators through other means.Federal Protective Service. Identify and locate potential criminal suspects using address, vehicle, and other information. Office of Detention and Removal.Locate and remove illegal aliens from the United States using address, vehicle, and other information.

Table 4. (Continued)

Major component	Information reseller	Aggregate contract value	Uses involving personal information
U.S. Customs and Border Protection	ChoicePoint, LexisNexis, Dun and Bradstreet, and West	$2,375,00 0	Conduct queries on people, businesses, property, and corresponding links in support of law enforcement, intelligence gathering, and prosecution support. Border Patrol Del Rio Sector.Obtain information such as addresses, telephone numbers, and names of relatives in support of investigations involving registered owners of seized vehicles and property. National Targeting Center. Look up information associated with license plate data to assist in vehicle identification at northern and southern landborders. License plate readers capture data on vehicles and cross-check against information reseller and government databases. Data capturedare used to help assess risk related to these vehicles (e.g., a car whose license plate is associated with a law enforcement record mightbe referred for secondary examination).
U.S. Citizenship and Immigration Services	ChoicePoint, LexisNexis, West	$960,000	Offices of Fraud Detection and National Security and Asylum.Detect fraud in applications for immigrant benefits and obtain court records (including judgments and conviction documents) to support a broad range of evidentiary requirements for official adjudication proceedings.
Transportation Security Administration	Acxiom, Insight America, Qsenta	$897,000	Test the feasibility of using commercial data sources to authenticate identity information contained in passenger records to support passenger prescreening.As part of the Secure Flight Program, TSA conducted a test to determine whether commercial data could be used to improve the effectiveness of watch list matching by identifying passengers who would not have been identified from passenger name records and government data alone. TSA plans to use the results of the test to identify what personally identifying information should be required in passenger name records to maximize aviation security.
U.S. Secret Service	ChoicePoint, Dallas Computer Services, Dun and Bradstreet, LocatePLUS, and APPRISS	$471,000	Provide investigative leads to field agents and other Secret Service personnel in conducting their investigations (e.g., to develop backgroundinformation on persons, locations, or businesses). Acquire jail data that are used as a cross-check against state and federal databases on warrants, sex offenders, child support, probations,and paroles.

Major component	Information reseller	Aggregate contract value	Uses involving personal information
Federal Emergency Management Agency	ChoicePoint	$113,000	Acquire information such as name, address, and Social Security numberto help verify and validate the identities of individuals applying for disaster assistance via the Internet.
Office of Inspector General	ChoicePoint, LexisNexis	$39,000	Generate leads in law enforcement investigations.
U.S. Coast Guard	ChoicePoint	$19,000	Obtain up-to-date credit reports as needed to assist in the resolution of financial issues that are of a security concern in adjudications.
Federal Law Enforcement Training Center— Special Investigations Division	ChoicePoint	$7,900	Verify addresses, conduct background checks, criminal and administrative investigations.

Source: DHS.

Notes: The table represents fiscal year 2005 contract values and may not reflect actual expenditures. We did not verify the accuracy or completeness of the dollar figures provided to us.

Contract values were rounded to the nearest thousand.

Several DHS components use the departmentwide contracts with LexisNexis and West. Components such as the Science and Technology and Management Directorates reported that their use of these departmentwide contracts did not involve the use of personal information (e.g., reported uses were for legal or scientific research); accordingly, we did not include these values in the table.

To the extent possible, we excluded uses that did not involve personal information; however, since DHS officials responsible for administering departmentwide FEDLINK contracts were unable to provide a breakdown of component billings by information reseller, the values reflected in the table may include uses that do not involve personal information. For example, U.S. Citizenship and Immigration Services' fiscal year 2005 use of departmentwide FEDLINK contracts totaled approximately $960,000, but contract officials could not provide specific amounts for this organization's use of ChoicePoint, LexisNexis, and West. Although U.S. Citizenship and Immigration Services described use of West as primarily for legal research, we could not separate costs associated with use of personal information.

[a]Acxiom, Insight America (now owned by Acxiom), and Qsent were subcontractors on the EagleForce Associates contract to conduct a commercial data test for the Secure Flight Program. Although EagleForce is not an information reseller, we included the contract value because the commercial data test involved the acquisition of personal information from resellers.

SSA Uses Information Resellers Primarily for Fraud Prevention and Identity Verification

In an effort to ensure the accuracy of Social Security benefit payments, SSA and its components reported using approximately $1.3 million in contracts in fiscal year 2005 with information resellers for a variety of purposes relating to fraud prevention (66 percent), such as skiptracing, [36] confirming suspected fraud related to workers compensation payments, obtaining information on criminal suspects for follow-up investigations (18 percent), and collecting debts (16 percent).

SSA and its components acquired these services through the use of the GSA and FEDLINK governmentwide contracts and their own contracts. In fiscal year 2005, SSA contracted with ChoicePoint, LexisNexis, SourceCorp, and Equifax.

The Office of the Inspector General (OIG), the largest user of information reseller data at SSA, supports the agency's efforts to prevent fraud, waste, and abuse. The OIG uses several information resellers to assist investigative agents in detecting benefit abuse by Social Security claimants and to assist agents in locating claimants. For example, OIG agents access reseller data to verify the identity of subjects undergoing criminal investigations.

Regional office agents may also use reseller data in investigating persons suspected of claiming disability fraudulently and draw upon assistance from OIG headquarters staff and state investigators from the state Attorney General's office in these investigations.

For example, the Northeastern Program Service Center, located in the New York branch of SSA, obtains New York State Workers Compensation Board data from SourceCorp, the only company legally permitted to maintain the physical and electronic records for New York State Workers Compensation. Through the use of this information, SSA can identify persons collecting workers compensation benefits but not reporting those benefits, as required, to the SSA.

Table 5 details in aggregate the vendors, fiscal year 2005 contract values, and uses of contracts with information resellers reported by major SSA components.

The Department of State Uses Information Resellers Primarily for Passport Fraud Detection and Investigation

The Department of State and its components reported approximately $569,000 in contracts in fiscal year 2005 with information resellers, primarily for assistance in fraud related activities through criminal investigations (51 percent), fraud detection (26 percent), and other uses (23 percent) such as background screening. State acquired information reseller services through the GSA schedule and a Justice blanket-purchase agreement. In fiscal year 2005, the majority of State contracts were with ChoicePoint; the agency also had contracts with LexisNexis, Equifax and Metronet.

State's components reported use of these contracts mainly for passport- related activities. For example, several components of State accessed personal information to validate information submitted on immigrant and nonimmigrant visa petitions, such as marital or familial relationships, birth and identity information, and address validation. A major use of reseller data at State is by investigators acquiring information on suspects in passport and visa fraud cases. According to State, information reseller data are increasingly important to its operations, because the number of passport and visa fraud cases has increased, and successful investigations of passport and visa fraud are critical to combating terrorism.

In addition to these uses, State acquires personal information through Equifax to support the financial background screening of its job applicants.

Table 6 details the vendors, fiscal year 2005 contract values, and uses of contracts with information resellers reported by major State components.

Table 5. Reported Uses of Personal Information: SSA Contracts with Information Resellers, Fiscal Year 2005

User	Information reseller	Contract value	Uses involving personal information
Agencywide	LexisNexis	$848,000a	Field Office Staff. Obtain resource information (i.e., real property ownership, values, real property transfers, and information concerning the ownership of automobiles and boats) to verify the validity of Supplemental Security Income applicants and recipients.Office of Inspector General. Access public records information to assistwith investigations of fraud and abuse within the SSA programs.Office of Hearings and Appeals. Access public records information to locate the addresses of individuals.
Office of the Inspector General	ChoicePoint	$240,000	Acquire information on subjects of criminal investigations (e.g., locations, assets, relatives) and help corroborate fraud allegations thatare submitted to the Office of the Inspector General by SSA or the general public.b
Agencywidec	Equifax	$204,000	Obtain address verification reports for the most current address of delinquent debtors for undeliverable overpayment-related notices and follow up billing and teleprinter profile reports (standard credit reports) that show the credit history of the debtor referred to Justice for enforcedcollection via civil suit.
Northeastern Program Service Center	SourceCorp	$14,000	Access New York State Worker Compensation Board payment data to ensure that persons claiming Social Security benefits are correctly reporting workers compensation benefits on their forms.

Source: SSA.

Notes: The table represents fiscal year 2005 contract values and may not reflect actual expenditures. We did not verify the accuracy or completeness of the dollar figures provided to us.

Contract values were rounded to the nearest thousand.

[a]This figure may include uses that do not involve personal information since LexisNexis provides news and legal research in addition to public records. SSA was unable to separate the dollar values associated with use of personal information from uses for other purposes.

[b]In addition to initiating its own investigations, the Office of the Inspector General receives notices from the general public about suspected fraud. According to one agency official, a large portion of these fraud allegations are either incomplete or unfounded and must be supported by substantial evidence. Before moving ahead with an investigation, officials obtain data from an information reseller to verify the legitimacy of the fraud allegations, fill in any missing information on the submitted forms and develop leads that would further the development of the allegation and any subsequent investigation if warranted.

[c]The Equifax data are accessible by the Northeastern Program Service Center, Mid-Atlantic Program Service Center, Southeastern Program Service Center, Great Lakes Program Service Center, Western Program Service Center, Mid-America Program Service Center, Office of Central Operations, and Office of Financial Policy and Operations.

[d]This is an SSA-funded joint investigation between SSA and the New Jersey State Attorney General's Office.

Table 6. Reported Uses of Personal Information: Department of State Contracts with Information Resellers, Fiscal Year 2005

Component	Information reseller	Contract value	Uses involving personal information
Diplomatic Security	ChoicePoint	$288,000	Criminal Investigations Division. Obtain leads on addresses, locations, identity, etc., used in the conduct of criminal investigations of passport and visa fraud. Diplomatic Security Command Centerand Diplomatic Security agents at26 overseas posts. Same use.
Office of Personnel Security and Suitability	Equifax	$132,000	Obtain credit checks on applicants and new hires to support backgroundscreening processes.
Bureau of Consular Affairs	ChoicePoint, Metronet	$89,000	Check the validity of selected passport applicatiqns, particularly two categories of high-risk applications.a
National Visa Center	ChoicePoint	$40,000	Verify information submitted on immigrant and nonimmigrant visa petitions.
Office of Consular Fraud Prevention Programs	LexisNexis	$21,000	Investigate claims of marital and familial relationships on immigrant visa applications and determine the bona fides of prospective employers for employment-based nonimmigrant visas.

Source: Department of State.

Note: The table represents fiscal year 2005 contract values and may not reflect actual expenditures. We did not verify the accuracy or completeness of the dollar figures provided to us.

[a]The two categories of high-risk passport applications include those with birth certificates from Puerto Rico and those from applicants lacking acceptable primary identification documents, who include affidavits from family or associates attesting to their identity.

Agencies Contract with Information Resellers Primarily through Use of GSA's Federal Supply Schedules and the Library of Congress's FEDLINK Service

In fiscal year 2005, the four agencies acquired personal information primarily through governmentwide contracts, including GSA's Federal Supply Schedule (52 percent) contracts and the Library of Congress's FEDLINK contracts (28 percent). Components within these agencies also initiated separate contracts with resellers as well. The Department of Justice was the largest user, accounting for approximately $19 million of the $30 million total for all four agencies. Figure 3 shows the values of reseller data acquisition by agency for fiscal year 2005.

In fiscal year 2005, the most common vehicles used among all four agencies to acquire personal information from information resellers were the governmentwide contracts made available through GSA's Federal Supply Schedule. The GSA schedule provides agencies with simplified, streamlined contracting vehicles, allowing them to obtain access to information resellers' services either by issuing task or purchase orders or by establishing blanket purchase agreements based on the schedule contracts. The majority of Justice's acquisition of information reseller services was obtained through the GSA schedule, including a blanket purchase agreement with ChoicePoint that was also made available to non- Justice agencies (for example, the Departments of State and Health and Human Services). In addition, components of DHS such as the U.S. Secret Service and the SSA's Office of Inspector General made use of GSA schedule contracts with information resellers.

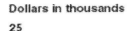

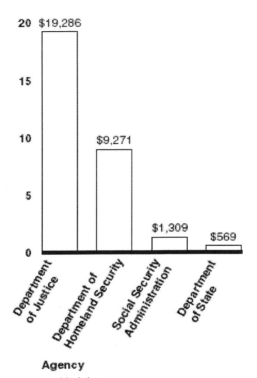

Source: GAO analysis of agency-provided data.

Figure 3. Total Dollar Values, Categorized by Agency, of Fiscal Year 2005 Acquisition of Personal Information from Information Resellers.

The Federal Supply Schedule allows agencies to take advantage of prenegotiated contracts with a variety of vendors, including information resellers. GSA does not assess fees for the use of these contracts; rather it funds the operation of the schedules in part by obtaining administrative fees from vendors on a quarterly basis. According to GSA officials, use of the schedule contracts allows agencies to obtain the best price and reduce their procurement lead time. Since these contracts have been prenegotiated, agencies do not need to issue their own solicitation. Instead, agencies may simply place a task order directly with the vendor, citing the schedule number. GSA's role in administering these contracts is primarily to negotiate baseline contract requirements and pricing; it does not monitor which agencies are using its schedule contracts. GSA officials noted that the requirements contained in the schedule contracts are baseline, and agencies may add more stringent requirements to their individual task orders.

Another contract vehicle commonly used to obtain personal information from information resellers was the Library of Congress's FEDLINK service (28 percent). This vehicle was used by both DHS and SSA. [37] FEDLINK, an intragovernmental revolving fund, [38] is a cooperative procurement, accounting, and training program designed to provide access to online databases, periodical subscriptions, books, and other library and information support services from commercial suppliers, including information resellers. At DHS, use of the FEDLINK service was the primary vehicle for contracting with information resellers. DHS also used GSA schedule buys, and some smaller purchases were made directly between DHS components and information resellers. The majority of SSA's fiscal year 2005 acquisitions from information resellers were through FEDLINK, with some use of the GSA schedule contracts.

FEDLINK allows agencies to take advantage of prenegotiated contracts at volume discounts with a variety of vendors, including information resellers. As with the GSA schedule contracts, the requirements of the FEDLINK contracts serve as a baseline, and agencies may add more stringent requirements if they so choose.

FEDLINK offers two different options for using its contracts: direct express and transfer pay. The direct express option is similar to the GSA schedule process, in which the agency issues a purchase order directly to the vendor and cites the underlying FEDLINK contract. Under direct express, the ordering agency is responsible for managing the delivery of products and services and paying invoices, and the vendor pays an administrative fee to the Library.

Under the transfer pay option, ordering agencies must sign an interagency agreement and pay an administrative fee to the Library. In turn, the ordering agencies receive additional administrative services. DHS used both the direct express and transfer pay options in fiscal year 2005, while SSA used transfer pay exclusively.

RESELLERS TAKE STEPS TO PROTECT PRIVACY, BUT THESE MEASURES ARE NOT FULLY CONSISTENT WITH THE FAIR INFORMATION PRACTICES

Although the information resellers that do business with the federal agencies we reviewed [39] have practices in place to protect privacy, these measures were not fully consistent with the Fair Information Practices. Most significantly, the first four principles, relating to collection limitation, data quality, purpose specification, and use limitation, are largely at odds with the nature of the information reseller business. These principles center on limiting the collection and use of personal information and require data accuracy based on that limited purpose and limited use of the information. However, the information reseller industry presupposes that the collection and use of personal information is not limited to specific purposes, but instead that information can be collected and made available to multiple customers for multiple purposes. Resellers make it their business to collect large amounts of personal information [40] and to combine that information in new ways so that it serves purposes other than those for which it was originally collected. Further, they are limited in their ability to ensure the accuracy, currency, or relevance of their holdings, because these qualities may vary based on customers' varying uses.

Information reseller policies and procedures were consistent with aspects of the remaining four Fair Information Practices. Large resellers reported implementing a variety of security safeguards, such as stringent customer credentialing, to improve protection of personal information. Resellers also generally provided public notice of key aspects of their privacy policies and practices, (relevant to the openness principle) and reported taking actions to ensure internal compliance with their own privacy policies (relevant to the accountability principle). However, resellers generally limited the extent to which individuals could gain access to personal information held about themselves, and because they obtain their information from other sources, most resellers also had limited provisions for correcting or deleting inaccurate information contained in their databases (relevant to the individual participation principle). [41] Instead, they directed individuals wishing to make corrections to contact the original sources of the data. Table 7 provides an overview of information resellers' application of the Fair Information Practices.

Information Resellers Generally Did Not Report Limiting Their Data Collection to Specific Purposes or Notifying Individuals about Them

According to the collection limitation principle of the Fair Information Practices, the collection of personal information should be limited, information should be obtained by lawful and fair means, and, where appropriate, it should be collected with the knowledge and consent of the individual. The collection limitation principle also suggests that organizations could limit collection to the minimum amount of data necessary to process a transaction.

Table 7. Information Resellers' Application of Principles of the Fair Information Practices

Principle	Resellers' application
Collection limitation. The collection of personal information should be limited, should be obtained by lawful and fair means, and, where appropriate, with the knowledge or consent of the individual.	Resellers do not limit collections to specific purposes but collectlarge amounts of personal information, within the bounds of thelaw. Further, in many cases, individuals do not know that their personal information is being collected by the reseller, even though they may have known of the original (source) collection.
Data quality. Personal information should be relevant to the purpose for which it is collected, and should be accurate, complete, and current as needed for that purpose.	Although they often have measures in place for ensuring data accuracy in the aggregate, resellers do not ensure that the information they provide is accurate, complete, and current for aspecific purpose. Instead, they monitor and rely on the quality controls of the original data source.
Purpose specification. The purpose for the collection of personal information should be disclosed before collection and upon any change to that purpose, and its use should be limited to that purpose and compatible purposes.	Resellers disclose general categories of purposes for their datacollection rather than specific purposes. They obtain informationoriginally collected for specific purposes and generally offer it fora much wider range of purposes.

Table 7 (Continued)

Principle	Resellers' application
Use limitation. Personal information should not be disclosed or otherwise used for other than a specified purpose without consent of the individual or legal authority.	Resellers generally limit the use of information as required by law rather than on the basis of the purposes originally specifiedwhen the information was collected. Resellers generally pass responsibility for legal use restrictions to customers through licensing and contract terms and agreements. Customers must contractually agree to appropriate uses of the data and must agree to comply with applicable laws.
Security safeguards. Personal information should be protected with reasonable security safeguards against risks such as loss or unauthorized access, destruction, use, modification, or disclosure.	Resellers reported implementing a variety of security safeguards, such as stringent customer credentialing, to improve protection of personal information.
Openness. The public should be informed about privacy policies and practices, and individuals should have ready means of learning about the use of personal information.	Resellers generally inform the public of key aspects of privacy policies through Web sites, brochures, and so on.
Individual participation. Individuals should have the following rights: to know about the collection of personal information, to access that information, to request correction, and to challenge the denial of those rights.	Although information resellers allow individuals access to their personal information, this access is generally limited, as is the opportunity to make corrections. Generally, resellers only correct errors they may have introduced in the process of obtaining and aggregating data.
Accountability. Individuals controlling the collection or use of personal information should be accountable for taking steps to ensure the implementation of these principles.	Resellers reported taking actions, such as designating a chief privacy officer or equivalent, to ensure compliance with their privacy policies. Annual privacy audits were conducted in one case.

Source: GAO analysis of reseller information. Note: We did not evaluate the effectiveness of information reseller practices, only the extent to which resellers applied the Fair Information Practices.

In practice, resellers are limited in the personal information that they can obtain by laws that apply to specific kinds of information (for example, the Fair Credit Reporting Act and the Gramm-Leach-Bliley Act, which restrict the collection, use, and disclosure of certain consumer and financial data). One reseller reported that it also restricts collection of Social Security number information from public records, as well as collection of identifying information on children from public sources, such as telephone directories.

Beyond specific legal restrictions, information resellers generally attempt to aggregate large amounts of personal information so as to provide useful information to a broad range of customers. For example, resellers collect personal information from a wide variety of sources, including state motor vehicle records; local government records on births, real property, and voter registrations; and various court records. Information resellers may also obtain information from telephone directories, Internet sites, and consumer applications for products or services. The widely varying sources and types of information demonstrate the broad nature of the collection of personal information. The amount and scope of information collected vary from company to company, and resellers use this information to offer a range of products tailored to different markets and uses. [42]

Regarding the principle that information should be obtained by lawful and fair means, resellers stated that they take steps to ensure that their collection of information is legal. For example, resellers told us that they obtain assurances from their data suppliers that information is legally collected from reputable sources. Further, they design their products and services to ensure they are in conformance with laws such as the GrammLeach-Bliley Act and the Fair Credit Reporting Act.

Regarding the principle that, where appropriate, information should be collected with the knowledge and consent of the individual, resellers do not make provisions to notify the individuals involved when they obtain personal data from their many sources, including public records. Concomitantly, individuals are not afforded an opportunity to express or withhold their consent when the information is collected. Resellers said they believe it may not be appropriate or practical for them to provide notice or obtain consent from individuals because they do not collect information directly from them. One reseller noted that in many instances the company does not have a direct relationship with the data subject and is therefore not in a position to interact with the consumer for purposes such as providing notice. Further, this reseller stated its belief that requiring resellers to notify and obtain consent from each individual about whom they obtain information would result in consumers being overwhelmed with notices and negate the value of notice.

Under certain conditions, some information resellers offer consumers an "opt-out" option—that is, individuals may request that information about themselves be suppressed from selected databases. However, resellers generally offer this option only with respect to certain types of information and only under limited circumstances. For example, one reseller allows consumers to opt out of its marketing products but not other products, such as background screening and fraud detection products. The privacy policy for another information reseller states that it will allow certain individuals to opt out of its nonpublic information databases containing sensitive information under specific conditions: if the individual is a state, local, or federal law enforcement officer or public official whose position exposes him or her to a threat of imminent harm; if the individual is a victim of identity theft; or if the individual is at risk of physical harm. In order to exercise this option, consumers generally must provide satisfactory documentation to support the basis for their request. In any event, the reseller retains the right to determine (1) whether to grant or deny any request, (2) to which databases the request for removal will apply, and (3) the duration of the removal. Two resellers stated their belief that under certain circumstances it may not be appropriate to provide consumers with opportunities for opting out, such as for information products designed to detect fraud or locate criminals. These resellers stated that if individuals were permitted to opt out of fraud prevention databases, some of those opting out could be criminals, which would undermine the effectiveness and utility of these databases.

Information Resellers Do Not Ensure That Personal Information They Provide Is Accurate for Specific Purposes

According to the data quality principle, personal information should be relevant to the purpose for which it is collected, and should be accurate, complete, and current as needed for that purpose. Information resellers reported taking steps to ensure that they generally receive accurate data from their sources and that they do not introduce errors in the process of transcribing and aggregating information; however, they generally provide their customers with exactly the same data they obtain and do not claim or guarantee that the information is accurate for a specific purpose. Some resellers' privacy policies state that they expect their data to contain some errors. Further, resellers varied in their policies regarding correction of data determined to be inaccurate as obtained by them. One reseller stated that it would delete information in its databases that was found to be inaccurate. Another stated that even if an individual presents persuasive evidence that certain information is in error, the reseller generally does not make changes if the information comes directly from an official public source (unless instructed to do so by that source). Because they are not the original source of the personal information, information resellers generally direct individuals to the original sources to correct any errors. Several resellers stated that they would correct any identified errors introduced through their own processing and aggregation of data.

While not providing specific assurance of the accuracy of the data they provide, information resellers reported that they take steps to ensure that their suppliers have data quality controls in place. For example, officials from one information reseller said they use a screening process to help determine whether they should use a particular supplier. [43] As part of this process, the reseller assesses whether the supplier has internal controls in place that are in line with the reseller's policies. Information resellers also reported that they conduct annual audits of their suppliers aimed at assessing the integrity and quality of the information they receive. If these audits show that a supplier has failed to provide accurate, complete, and timely information, the reseller may discontinue using that supplier.

Resellers also noted that data accuracy is contingent upon intended use. That is, data that may be perfectly adequate for one purpose may not be precise enough or appropriate for another purpose. While end users, such as federal agencies, may address data quality for their specific purposes, resellers—who maintain personal information for multiple purposes—are less able to achieve accuracy because they support multiple uses. Thus, resellers generally disclaim data accuracy and leave it to their customers to ensure that the data are accurate for their intended uses. One reseller stated that their customers understand the accuracy limitations of the data they obtain and take the potential for data inaccuracy into account when using the data.

Information Resellers' Specification of the Purpose of Data Collection Consists of Broad Descriptions of Business Categories

According to the purpose specification principle, the purpose for the collection of personal information should be disclosed before collection and upon any change to that purpose, and its use should be limited to that purpose and compatible purposes. While information resellers specify purpose in a general way by describing the types of businesses that use their data, they generally do not designate specific intended uses for each of their data collections. Resellers generally obtain information that has already been collected for a specific purpose and make that information available to their customers, who in turn have a broader variety of purposes for using it. For example, personal information originally submitted by a customer to register a product warranty could be obtained by a reseller and subsequently made available to another business or government agency, which might use it for an unrelated purpose, such as identity verification, background checking, or marketing.

In a general sense, information resellers specify their purpose by indicating (on company Web sites, for example) the business categories of the customers for whom they collect information. For example, reseller privacy policies generally state that resellers make personal information available for legitimate uses by business and government organizations. Examples of business categories may be provided, but resellers do not specify which types of information are to be used in which business categories. It is difficult for resellers to provide greater specificity because they make their data available to many customers for a wide range of legitimate purposes. As a result, the public is made aware only of the broad range of potential uses to which their personal information may be applied, rather than a specific use, as envisioned in the Fair Information Practices.

Information Resellers Generally Limit the Use of Information as Required by Law, Rather Than on the Basis of Purposes Originally Specified When the Information Was Collected

Under the use limitation principle, personal information should not be disclosed or used for other than the originally specified purpose without consent of the individual or legal authority. However, because information reseller purposes are specified very broadly, it is difficult for resellers to ensure that use of the information in their databases is limited. As previously discussed, information reseller data may have many different uses, depending on the types of customers involved. Resellers do take steps to ensure that their customers' use of personal information is limited to legally sanctioned purposes. Information resellers pass this responsibility to their customers through licensing agreements and contract terms and agreements.

According to two large information resellers, customers are generally contractually required to use data from resellers appropriately and must agree to comply with applicable laws, such as the Gramm-Leach-Bliley Act, the Fair Credit Reporting Act, and the Driver's Privacy Protection Act. For example, one information reseller uses a service agreement that includes provisions governing permissible use of information sought by the customer, the confidentiality of information provided, legal requirements under federal and state laws, and other customer obligations. The reseller reported that the company monitors its customers' compliance by conducting periodic audits and taking appropriate actions in response to any audit findings.

In a standardized agreement form used by another reseller, federal agencies must certify that they will use information obtained from the reseller only as permissible under the Gramm-Leach-Bliley Act and the Driver's Privacy Protection Act. The service agreement identifies permissible purposes for information whose use is restricted by these laws and requires agencies to agree that they will use the information only in the performance or the furtherance of appropriate government activities. In conformance with the Gramm-Leach-Bliley Act permissible uses, the information reseller requires agencies to certify that they will use personal information "only as requested or authorized by the consumer."

The information resellers used by the federal agencies we reviewed generally also reported taking steps to ensure that access to certain sensitive types of personally identifiable information is limited to certain customers and uses. For example, two resellers reported that they provide full Social Security numbers and driver's license numbers only to specific types of customers, including law enforcement agencies and insurance companies, and for purposes such as employment or tenant screening. While actions such as these are useful in protecting privacy and are consistent with the use limitation principle in that they narrow the range of potential uses for this type of information, they are not equivalent to limiting use only to a specific predefined purpose. Without limiting use to predefined purposes, resellers cannot provide individuals with assurance that their information will only be accessed and used for the purpose originally specified when the information was collected.

Information Resellers Reported Taking Steps to Improve Security Safeguards

According to the security safeguards principle, personal information should be protected with reasonable safeguards against risks such as loss or unauthorized access, destruction, use, modification, or disclosure. While we did not evaluate the effectiveness of resellers' information security programs, resellers we spoke with said they employ various safeguards to protect consumers' personal information. They implemented these safeguards in part for business reasons but also because federal laws require such protections. Resellers describe these safeguards in various policy statements, such as online and data privacy policies or privacy statements posted on Internet sites. Resellers also generally had information security plans describing, among other things, access controls for information and systems, document management practices, incident reporting, and premises security.

Given recent incidents, large information resellers reported having recently taken steps to improve their safeguards against unauthorized access. In a well-publicized incident, in February 2005, ChoicePoint disclosed that unauthorized individuals had gained access to personal information by posing as a firm of private investigators. In the following month, LexisNexis disclosed that unauthorized individuals had gained access to personal information through the misappropriation of user IDs and passwords from legitimate customers. These disclosures were required by state law, as previously discussed. In January 2006, ChoicePoint reached a settlement with the Federal Trade Commission [44] over charges that the company did not have reasonable procedures to verify the identity of prospective new users. The company agreed to implement new procedures to ensure that it provides consumer reports only to legitimate business for lawful purposes. In the mean time, both information resellers reported that they had taken steps to improve their procedures for authorizing customers to have access to sensitive information, such as Social Security numbers. For example, one reseller established a credentialing task force with the goal of centralizing its customer credentialing process. In order for customers of this reseller to obtain products and services containing sensitive personal information, they must now undergo a credentialing process involving a site visit by the information reseller to verify the accuracy of information reported about the business. Applicants are then scored against a credentialing checklist to determine whether they will be granted access to sensitive information. In addition, both resellers reported efforts to strengthen user ID and password protections and restrict access to sensitive personal information (including full driver's license numbers and Social Security numbers) to a limited number of customers, such as law enforcement agencies (others would be able to view masked information). Although we did not test the effectiveness of these measures, if implemented correctly, they could help provide assurance that sensitive information is protected appropriately.

In addition to enhancing safeguards on customer access authorizations, resellers have instituted a variety of other security controls. For example, three large information resellers have implemented physical safeguards at their data centers, such as continuous monitoring of employees entering and exiting facilities, monitoring of activity on customer accounts, and strong authentication of users entering and exiting secure areas within the data centers. Officials at one reseller told us that security profiles were established for each employee that restrict access to various sections of the center based upon employee job functions. Computer rooms were further protected with a combined system of biometric hand readers and security codes. Security cameras were placed throughout the facility for continuous recording of activity and review by security staff. Information resellers also had contingency plans in place to continue or resume operations in the event of an emergency.

Information resellers reported that on an annual basis, or more frequently if needed, they conduct security risk assessments as well as internal and external security audits. These assessments address such topics as vulnerabilities to internal or external security threats, reporting and responding to security incidents, controls for network and physical facilities, and business continuity management. The assessments also addressed strategies for mitigating potential or identified risks.

If properly implemented, security measures such as those reported by information resellers could contribute to effective implementation of the security safeguards principle.

Information Resellers Generally Informed the Public about Their Privacy Policies and Practices

According to the openness principle, the public should be informed about an organization's privacy policies and practices, and individuals should have ready means of learning about the organization's use of personal information.

To address openness, information resellers took steps to inform the public about key aspects of their privacy policies. They used means such as company Web sites and brochures to inform the public of specific policies and practices regarding the collection and use of personal information. Reseller Web sites also generally provided information about the types of information products the resellers offered—including product samples—as well as general descriptions about the types of customers served. Several Web sites also provided advice to consumers on protecting personal information and discussed what to do if individuals suspect they are victims of identity theft.

Providing public notice of privacy policies informs individuals of what steps an organization takes to protect the privacy of the personal information it collects and helps to ensure the organization's accountability for its stated policies.

Information Reseller Policies Generally Allow Individuals Limited Ability to Access and Correct Their Personal Information

According to the individual participation principle, individuals should have the right to know about the collection of personal information, to access that information, to request correction, and to challenge the denial of those rights. Information resellers generally allow individuals access to their personal information. However, this access is limited, as is the opportunity to make corrections. Resellers may provide an individual a report containing certain types of information—such as compilations of public records information—however, the report may not include all information maintained by the resellers about that individual. For example, one information reseller stated that it offers a free report, under certain circumstances, on an individual's claims history, employment history, or tenant history. Resellers may offer basic reports to individuals at no cost, but they generally charge for reports on additional information. A free consumer report, such as an employment history report, for example, typically excludes information such as driver's license data, family information, and credit header data that a reseller may possess in other databases.

Although individuals can access information about themselves, if they find inaccuracies, they generally cannot have these corrected by the resellers. [45] Information resellers direct individuals to take their cases to the original data sources—such as courthouses or other local government agencies— and attempt to have the inaccuracy corrected there. Several resellers stated that they would correct any identified errors introduced through their own processing and aggregation of data. As discussed above, resellers, as a matter of policy, do not make corrections to data obtained from other sources, even if the consumer provides evidence that the data are wrong.

According to resellers, making corrections to their own databases is extremely difficult, for several reasons. First, the services these resellers provide concentrate on providing references to a particular individual from many sources, rather than distilling only the most accurate or current reference. For example, a reseller might have many instances in its databases of a particular individual's current address. Although most might be the same, there could be errors as well. Resellers generally would report the information as they have it rather than attempting to determine which entry is correct. This information is important to customers such as law enforcement agencies. Further, resellers stated that making corrections to their databases could be ineffective because the data are continually refreshed with updated data from the source, and thus any correction is likely to be changed back to its original state the next time the data are updated. In addition, as discussed in the collection limitation section, resellers stated their belief that it would not be appropriate to allow the public to access and correct information held for certain purposes, such as fraud detection and locating criminals, since providing such rights could undermine the effectiveness of these uses (e.g., by allowing criminals to access and change their information). However, as a result of these practices, individuals cannot know the full extent of personal information maintained by resellers or ensure its accuracy.

Information Resellers Report Measures to Ensure Accountability for the Collection and Use of Personal Information

According to the accountability principle, individuals controlling the collection or use of personal information should be accountable for taking steps to ensure the implementation of the Fair Information Practices. Although information resellers' overall application of the Fair Information Practices varied, each reseller we spoke with reported actions to ensure compliance with its own privacy policies. For example, resellers reported designating chief privacy officers to monitor compliance with internal privacy policies and applicable laws (e.g., the Gramm-Leach-Bliley Act and the Driver's Privacy Protection Act). Information resellers reported that these officials had a range of responsibilities aimed at ensuring accountability for privacy policies, such as establishing consumer access and customer credentialing procedures, monitoring compliance with federal and state laws, and evaluating new sources of data (e.g., cell phone records).

Auditing of an organization's practices is one way of ensuring accountability for adhering to privacy policies and procedures. Although there are no industrywide standards requiring resellers to conduct periodic audits of their compliance with privacy policies, one information reseller reported using a third party to conduct privacy audits on an annual basis. Using a third party to audit compliance with privacy policies further helps to ensure that an information reseller is accountable for the implementation of its privacy practices.

Establishing accountability is critical to the protection of privacy. Actions taken by data resellers should help ensure that their privacy policies are appropriately implemented.

AGENCIES LACK POLICIES ON USE OF RESELLER DATA, AND PRACTICES DO NOT CONSISTENTLY REFLECT THE FAIR INFORMATION PRACTICES

Agency practices for handling personal information acquired from information resellers did not always fully reflect the Fair Information Practices. Further, agencies generally lacked policies that specifically address their use of personal information from commercial sources, although DHS Privacy Office officials reported that they were drafting such a policy.

As shown in table 8, four of the Fair Information Practices—the collection limitation, data quality, use limitation, and security safeguards principles—were generally reflected in agency practices. For example, several agency components (specifically, law enforcement agencies such as the FBI and the U.S. Secret Service) reported that in practice, they generally corroborate information obtained from resellers when it is used as part of an investigation.

Table 8. Application of Fair Information Practices to the Reported Handling of Personal Information from Data Resellers at Four Agencies

Principle	Agency application of principle	Agency practices
Collection limitation. The collection of personal information should be limited, should be obtained by lawful and fair means, and, where appropriate, with the knowledge or consent of the individual.	General	Agencies limited personal data collection to individuals under investigation or their associates.
Data quality. Personal information should be relevant to the purpose for which it is collected, and should be accurate, complete, and current as needed for that purpose.	General	Agencies corroborated information from resellers and did not take actions based exclusively on such information.
Purpose specification. The purpose for the collection of personal information should be disclosed before collection and upon any change to that purpose, and its use should be limited to that purpose and compatible purposes.	Uneven	Agency system of records notices did not generally reveal that agency systems could incorporate information from data resellers. Agencies also generally did not conduct privacy impact assessments for their systems or programs that involve use of reseller data.
Use limitation. Personal information should not be disclosed or otherwise used for other than a specified purpose without consent of the individual or legal authority.	General	Agencies generally limited their use of personal information to specific investigations (including law enforcement, counterterrorism, fraud detection, and debt collection).
Security safeguards. Personal information should be protected with reasonable security safeguards against risks such as loss or unauthorized access, destruction, use, modification, or disclosure.	General	Agencies had security safeguards such as requiring passwordsto access databases, basing access rights on need to know, and logging search activities (including "cloaked logging," whichprevents the vendor from monitoring search content).
Openness. The public should be informed about privacy policies and practices, and individuals should have ready means of learning about the use of personal information.	Uneven	See Purpose specificationabove. Agencies did not have established policies specifically addressing the use of personal information obtained from resellers.
Individual participation. Individuals should have the following rights: to know about the collection of personal information, to access that information, to request correction, and to challenge the denial of those rights.	Uneven	See Purpose specificationabove. Because agencies generally did not disclose their collections of personal information from resellers, individuals were often unable to exercise these rights.

Principle	Agency application of principle	Agency practices
Accountability. Individuals controlling the collection or use of personal information should be accountable for taking steps to ensure the implementation of these principles.	Uneven	Agencies do not generally monitor usage of personal information from information resellers to hold users accountable for appropriate use; instead, they rely on users to be responsible for their behavior. For example, agencies may instruct users in their responsibilities to use personal information appropriately, have them sign statements of responsibility, and have them indicate what permissible purpose a given search fulfills.

Legend:

General = policies or procedures to address all major aspects of a particular principle.

Uneven = policies or procedures addressed some but not all aspects of a particular principle or some but not all agencies and components had policies or practices in place addressing the principle.

Source: GAO analysis of agency-supplied data.

Note: We did not independently assess the effectiveness of agency information security programs. Our assessment of overall agency application of the Fair Information Practices was based on the policies and management practices described by the Department State and SSA as a whole and by major components of Justice and DHS (footnote 2 in app. I lists these components). We did not obtain information on smaller components of Justice and DHS.

This practice is consistent with the data quality principle that data should be accurate, current, and complete. Agency policies and practices with regard to the other four principles, however, were uneven. Specifically, agencies did not always have policies or practices in place to address the purpose specification, openness, and individual participation principles with respect to reseller data. The inconsistencies in application of these principles as well as the lack of specific agency policies can be attributed in part to ambiguities in OMB guidance regarding the applicability of the Privacy Act to information obtained from resellers. Further, privacy impact assessments, which often are not conducted, are a valuable tool that could address important aspects of the Fair Information Practices. Finally, components within each of the four agencies did not consistently hold staff accountable by monitoring usage of personal information from information resellers and ensuring that it was appropriate; thus, their application of the accountability principle was uneven.

Agency Procedures Reflect the Collection Limitation, Data Quality, Use Limitation, and Security Safeguards Principles

The collection limitation principle establishes, among other things, that organizations should obtain only the minimum amount of personal data necessary to process a transaction. This principle also underlies the Privacy Act requirement that agencies maintain in their records "only such information about an individual as is relevant and necessary to accomplish a purpose of the agency." [46] Regarding most law-enforcement and counterterrorism purposes, which accounted for 90 percent of usage in fiscal year 2005, agencies generally limited their personal data collection in that they reported obtaining information only on specific individuals under investigation or associates of those individuals.47 Having initiated investigations on specific individuals, however, agencies generally reported that they obtained as much personal information as possible about the individuals being investigated, because law enforcement investigations require pursuing as many investigative leads as possible.

The data quality principle states that, among other things, personal information should be relevant to the purpose for which it is collected and be accurate. This principle is mirrored in the Privacy Act's requirement for agencies to maintain all records used to make determinations about an individual with sufficient accuracy, relevance, timeliness, and completeness as is reasonably necessary to ensure fairness. [48]

Agencies reported taking steps to mitigate the risk of inaccurate information reseller data by corroborating information obtained from resellers. Agency officials described the practice of corroborating information as a standard element of conducting investigations. Officials from several law enforcement component agencies, including ATF and DEA, said corroboration was necessary to build legally sound cases from investigations. For example, U.S. Secret Service officials reported that they instruct agents that the information obtained from resellers should be independently corroborated, and that none of it should be used as probable cause for obtaining warrants.

Further, FBI officials from FTTTF noted that obtaining data from information resellers helps to improve the overall quality and accuracy of the data in investigative files. Officials stated that the variety of private companies providing personal information enhances the value, quality, and diversity of the information used by the FBI, noting that a decision to put an individual under arrest is based on "probable cause," which is determined by a preponderance of evidence, rather than any single source of information, such as information in a reseller's data base.

Likewise, for non law-enforcement use, such as debt collection and fraud detection and prevention, agency components reported procedures for mitigating potential problems with the accuracy of data provided by resellers by obtaining additional information from other sources when necessary. For example, the Executive Office for U.S. Attorneys uses information resellers to obtain information on assets possessed by an individual indebted to the United States. According to officials, should information contained in the information reseller databases conflict with informataion provided by an individual, further investigation takes place before any action to collect debts would be taken. Likewise, officials from the U.S. Citizenship and Immigration Services (USCIS) component of DHS and the Office of Consular Affairs within the Department of State reported similar practices. While these practices do not eliminate inaccuracies in data coming into the agency, they help ensure the quality of the information that is the basis for agency actions.

The use limitation principle provides that personal information should not be disclosed or used for other than a specified purpose without consent of the individual or legal authority. This principle underlies the Privacy Act requirement that prevents agencies from disclosing records on individuals except with consent of the individual, unless disclosure of the record would be, for example, to another agency for civil or criminal law enforcement activity or for a purpose that is compatible with the purpose for which the information was collected [49].

Although agencies rely on resellers' multipurpose collection of information as a source, agency officials said their use of reseller information was limited to distinct purposes, which were generally related to law enforcement or counterterrorism. For example, the Department of Justice reported uses specific to the conduct of criminal investigations on individuals, terrorism investigations, and the location of assets and witnesses. Other Justice and DHS components, such as the Federal Protective Service, U.S. Secret Service, FBI, and ATF, also reported that they used information reseller data for investigations. For uses not related to law enforcement, such as those reported by State and SSA, use of reseller information was also described as supporting a specific purpose (e.g., fraud detection or debt collection).

The use limitation principle also precludes agencies from sharing personal information they collect for purposes unrelated to the original intended use of the information. Officials of certain law enforcement components of these agencies reported that in certain cases they share information with other law enforcement agencies, a use consistent with the purposes originally specified by the agency. For example, the FBI's FTTTF supports ongoing investigations in other law enforcement agencies and the intelligence community by sharing information obtained from resellers (among other information) in response to requests about foreign terrorists from FBI agents or officials from partner agencies. [50]

The security safeguards principle requires that personal information be reasonably protected against unauthorized access, use, or disclosure. This principle also underlies the Privacy Act requirement that agencies establish appropriate administrative, technical, and physical safeguards to ensure the security and confidentiality of records on individuals. [51] This principle is further mirrored in the FISMA requirement to protect information and information systems from unauthorized access, use, disclosure, disruption, modification, or destruction, including through controls for confidentiality.

While we did not assess the effectiveness of information security or the implementation of FISMA at any of these agencies, we found that all four had measures in place intended to safeguard the security of personal information obtained from resellers. [52] For example, all four agencies cited the use of passwords to prevent unauthorized access to information reseller databases. Further, agency components such as ATF, DEA, CBP, and USCIS, reported that they limit access to sensitive personal information (e.g., full Social Security number, driver's license number) to those with a specific need for this information. Several agency components also reported that resellers were promptly notified to deactivate accounts for employees separated from government service to protect against unauthorized use. As another security measure, several components, including DEA and the FBI, reported that resellers notified them when accounts were accessed from Internet addresses at unexpected locations, such as outside the United States.

Another measure to prevent unauthorized disclosure reported by law enforcement agencies, such as the FBI, ICE, and Secret Service, is the use of "cloaked logging," which prevents vendor personnel from monitoring the queries being made by law enforcement agents. Officials in FBI's FTTTF reported that, in order to maintain the integrity of investigations, resellers are contractually prohibited from tracking or monitoring the exact persons or other entities being searched by FTTTF personnel. Law enforcement officials stated that the ability to mask searches from vendors is important so that those outside law enforcement have no knowledge of who is being investigated and so that subjects of an investigation are not "tipped off."

Agency adherence to the collection limitation, data quality, use limitation, and security safeguards principles was based on general business procedures—including law-enforcement investigative practices— that reflect security and civil liberties protections, rather than written policies specifically regarding the collection, accuracy, use, and security of personal information obtained from resellers. Implementation of these practices provides individuals with assurances that only a limited amount of their personal information is being collected, that it is used only for specific purposes, and that measures are in place to corroborate the accuracy of the information and safeguard it from improper disclosure. These controls help prevent potential harm to individuals and invasion of their privacy by limiting the exposure of their information and reducing the likelihood of inaccurate data being used to make decisions that could affect their welfare.

Limitations in the Applicability of the Privacy Act and Ambiguities in OMB Guidance Contribute to an Uneven Adherence to the Purpose Specification, Openness, and Individual Participation Principles

The purpose specification, openness, and individual participation principles stipulate, among other things, that individuals should be made aware of the purpose and intended uses of the personal information being collected about them and have the ability to access and correct such information, if necessary. The Privacy Act reflects these principles in part by requiring agencies to publish in the Federal Register, "upon establishment or revision, a notice of the existence and character of a system of records." This notice is to include, among other things, the categories of records in the system as well as the categories of sources of records. [53]

In a number of cases, agencies did not adhere to the purpose specification or openness principles in regard to their use of reseller information in that they did not notify the public that they were using such information and did not specify the purpose for their data collections. Agency officials said that they generally did not prepare system-of-records notices that would address these principles because they were not required to do so by the Privacy Act. The act's vehicle for public notification—the system-ofrecords notice—becomes binding on an agency only when the agency collects, maintains, and retrieves personal data in the way defined by the act or when a contractor does the same thing explicitly on behalf of the government. Agencies generally did not issue system-of-records notices specifically for their use of information resellers largely because information reseller databases were not considered "systems of records operated by or on behalf of a government agency" and thus were not considered subject to the provisions of the Privacy Act. [54] OMB guidance on implementing the Privacy Act does not specifically refer to the use of reseller data or how it should be treated. According to OMB and other agency officials, information resellers operate their databases for multiple customers, and federal agency use of these databases does not amount to the operation of a system of records on behalf of the government. Further, agency officials stated that merely querying information reseller databases did not amount to agency "maintenance" of the personal information being queried and thus also did not trigger the provisions of the Privacy Act. In many cases, agency officials considered their use of resellers to be of this type—essentially "ad hoc" querying or "pinging" of reseller databases for personal information about specific individuals, which they believed they were not doing in connection with a formal system of records.

In other cases, however, agencies maintained information reseller data in systems for which system-of-records notices had been previously published. For example, law enforcement agency officials stated that, to the extent they retain the results of reseller data queries, this collection and use is covered by the system of records notices for their case file systems. However, in preparing such notices, agencies generally did not specify that they were obtaining information from resellers. Among system of records notices that were identified by agency officials as applying to the use of reseller data, only one—TSA's system of records notice for the test phase of its Secure Flight program—specifically identified the use of information reseller data. [55] Other programs that involve use of information reseller data include the fraud prevention and detection programs reported by SSA and State as well as law enforcement programs within ATF, the U.S. Marshals, and USCIS. For these programs, associated system of records notices identified by officials did not specify the use of information reseller data.

In several of these cases, agency sources for personal information were described only in vague terms, such as "private organizations," "other public sources," or "public source material," when information was being obtained from information resellers. [56] In one case, a notice indicated incorrectly that personal information was collected only from the individuals concerned. Specifically, USCIS prepared a system of records notice covering the Computer Linked Application Information Management System, which did not identify information resellers as a source. Instead, the notice stated only that "information contained in the system of records is obtained from individuals covered by the system." [57]

The inconsistency with which agencies specify resellers as a source of information in system-of-records notices is in part due to ambiguity in OMB guidance, which states that "for systems of records which contain information obtained from sources other than the individual to whom the records pertain, the notice should list the types of sources used." Although the guidance is unclear what would constitute adequate disclosure of "types of sources," OMB and DHS Privacy Office officials agreed that to the extent that reseller data are subject to the Privacy Act, agencies should specifically identify information resellers as a source and that merely citing public records information does not sufficiently describe the source.

The individual participation principle gives individuals the right to access and correct information that is maintained about them. However, under the Privacy Act, agencies can claim exemptions from the requirement to provide individual access and the ability to make corrections if the systems are for law enforcement purposes. [58] In most cases where officials identified system-of-record notices associated with reseller data collection for law enforcement purposes, agencies claimed this exemption. Like the ability to mask database searches from vendors, this provision is important so that the subjects of law enforcement investigations are not tipped off.

Aside from the law enforcement exemptions to the Privacy Act, adherence to the purpose specification and openness principles is critical to preserving a measure of individual control over the use of personal information. Without clear guidance from OMB or specific policies in place, agencies have not consistently reflected these principles in their collection and use of reseller information. As a result, without being notified of the existence of an agency's information collection activities, individuals have no ability to know that their personal information could be obtained from commercial sources and potentially used as a basis, or partial basis, for taking action that could have consequences for their welfare.

Privacy Impact Assessments Could Address Openness, and Purpose Specification Principles but Are Often Not Conducted

The PIA is an important tool for agencies to address privacy early in the process of developing new information systems, and to the extent that PIAs are made publicly available, [59] they provide explanations to the public about such things as the information that will be collected, why it is being collected, how it is to be used, and how the system and data will be maintained and protected. In doing so, they serve to address the openness and purpose specification principles.

However, only three agency components reported developing PIAs for their systems or programs that make use of information reseller data. [60] As with system-of-records notices, agencies often did not conduct PIAs because officials did not believe they were required.

Current OMB guidance on conducting PIAs is not always clear about when they should be conducted. According to guidance from OMB, a PIA is required by the E-Government Act when agencies "systematically incorporate into existing information systems databases of information in identifiable form purchased or obtained from commercial or public sources." [61] However, the same guidance also instructs agencies that "merely querying a database on an ad-hoc basis does not trigger the PIA requirement." Reported uses of reseller data were generally not described as a "systematic" incorporation of data into existing information systems; rather, most involved querying a database and in some cases retaining the results of these queries. OMB officials stated that agencies would need to make their own judgments on whether retaining the results of searches of information reseller databases constituted a "systematic incorporation" of information.

DHS has recently developed guidance requiring PIAs to be conducted whenever reseller data are involved. The DHS Privacy Office [62] guidance on conducting PIAs points out, for example, that a program decision to obtain information from a reseller would constitute a new source of information, requiring that a PIA be conducted. However, although the DHS guidance clearly states that PIAs are required when personally identifiable information is obtained from a commercial source, it also states that "merely querying such a source on an ad hoc basis using existing technology does not trigger the PIA requirement." [63] Like OMB's guidance, the DHS guidance is not clear, because agency personnel are left to make individual determinations as to whether queries are "on an ad hoc basis."

In one case, a DHS component prepared a PIA for a system that collects reseller data but had not identified in the assessment that resellers were being used. DHS's USCIS uses copies of court records obtained from an information reseller to support evidentiary requirements for official adjudication proceedings concerning fraud. Although this use was reported to be covered by the PIA for the office's Fraud Tracking System, the PIA identifies only "public records" as the source of its information and does not mention that the public records are obtained from information resellers. [64] In contrast, the draft DHS guidance on PIAs instructs DHS component agencies to "list the individual, entity, or entities providing the specific information identified above. For example, is the information collected directly from the individual as part of an application for a benefit, or is it collected from another source such as a commercial data aggregator." At the time of our review, this draft guidance had not yet been disseminated to DHS components. Lacking such guidance, DHS components did not have policies in place regarding the conduct of PIAs with respect to reseller data, nor did other agencies we reviewed.

Until PIAs are conducted more thoroughly and consistently, the public is likely to remain incompletely informed about agency purposes and uses for obtaining reseller information.

Agencies Often Did Not Have Practices in Place to Ensure Accountability for Proper Handling of Information Reseller Data

According to the accountability principle (individuals controlling the collection or use of personal information should be accountable for taking steps to ensure the implementation of the Fair Information Practices), agencies should take steps to ensure that employee uses of personal information from information resellers are appropriate. While agencies described activities to oversee the use of information resellers, such activities were largely based on trust of the user to use the information appropriately. For example, in describing controls placed on the use of commercial data, officials from component agencies identified measures such as instructing users that reseller data are for official use only and requiring users to sign statements of responsibility attesting to a need to access the information reseller databases and that their use will be limited to official business. Additionally, agency officials reported that in accessing reseller databases, users are required to select from a list of vendor-defined "permissible purposes" (e.g., law enforcement, transactions authorized by the consumer) before conducting a search. While these practices appear consistent with the accountability principle, they are focused on individual user responsibility rather than management oversight.

For example, agencies did not have practices in place to obtain reports from resellers that would allow them to monitor usage of reseller databases at a detailed level. Although agencies generally receive usage reports from the information resellers, these reports are designed primarily for monitoring costs. Further, these reports generally contained only high-level statistics on the number of searches and databases accessed, not the contents of what was actually searched, thus limiting their utility in monitoring usage. For example, one information reseller reported that it does not provide reports to agencies on the "permissible purpose" that a user selects before conducting a search.

Not all component agencies lacked robust user monitoring. Specifically, according to FBI officials from the FTTTF, their network records and monitors searches conducted by the user account, including who is searched against what public source database. The system also tracks the date and time of the query as well as what the analyst does with the data. FBI officials stated that the vendor reports as well as the network monitoring provide FBI with the ability to detect unusual usage of the public source providers.

To the extent that federal agencies do not implement methods such as user monitoring or auditing of usage records, they provide limited accountability for their usage of information reseller data and have limited assurance that the information is being used appropriately.

CONCLUSIONS

Services provided by information resellers serve as important tools that can enhance federal agency functions, such as law enforcement and fraud protection and identification.

Resellers have practices in place to protect privacy, but these practices are not fully consistent with the Fair Information Practices. Among other things, resellers collect large amounts of information about individuals without their knowledge or consent, do not ensure that the data they make available are accurate for a given purpose, and generally do not make corrections to the data when errors are identified by individuals. Information resellers believe that application of the relevant principles of the Fair Information Practices is inappropriate or impractical in these situations. Given that reseller data may be used for a variety of purposes, determining the appropriate degree of control or influence individuals should have over the way in which their personal information is obtained and used—as envisioned in the Fair Information Practices—is critical. To more fully embrace these principles could require resellers to change the way they conduct business, and currently resellers are not legally required to follow them. As Congress weighs various legislative options, adherence to the Fair Information Practices will be an important consideration in determining the appropriate balance between the services provided by information resellers to customers such as government agencies and the public's right to privacy.

Agencies take steps to adhere to Fair Information Practices such as the collection limitation, data quality, use limitation, and security safeguards principles. However, they have not taken all the steps they could to reflect others—or to comply with specific Privacy Act and e- Government Act requirements—in their handling of reseller data. Specifically, agencies did not always have policies or practices in place to address the purpose specification, individual participation, openness, and accountability principles with respect to reseller data. An important factor contributing to this is that OMB privacy guidance does not clearly address information reseller data, which has become such a valuable and useful tool for agencies. As a result, agencies are left largely on their own to determine how to satisfy legal requirements and protect privacy when acquiring and using reseller data. Without current and specific guidance, the government risks continued uneven adherence to important, well- established privacy principles and lacks assurance that the privacy rights of individuals are adequately protected.

MATTER FOR CONGRESSIONAL CONSIDERATION

In considering legislation to address privacy concerns related to the information reseller industry, Congress should consider the extent to which the industry should adhere to the Fair Information Practices.

RECOMMENDATIONS FOR EXECUTIVE ACTION

To improve accountability, ensure adequate public notice of agencies' use of personal information from commercial sources, and allay potential privacy concerns arising from agency use of information from such sources, we are making three recommendations to the Director of OMB and the heads of the four agencies. Specifically, we recommend that:

- the Director of OMB revise guidance on system of records notices and privacy impact assessments to clarify the applicability of the governing laws (the Privacy Act and the E-Government Act) to the use of personal information from resellers. These clarifications should specify the circumstances under which agencies should make disclosures about their uses of reseller data so that agencies can properly notify the public (for example, what constitutes a "systematic" incorporation of reseller data into a federal system). The guidance should include practical scenarios based on uses agencies are making of personal information from information resellers (for example, visa, criminal, and fraud investigations).
- the Director of OMB direct agencies to review their uses of personal information from information resellers, as well as any associated system of records notices and privacy impact assessments, to ensure that such notices and assessments explicitly reference agency use of information resellers.
- the Attorney General, the Secretary of Homeland Security, the Secretary of State, and the Commissioner of SSA develop specific policies for the collection, maintenance, and use of personal information obtained from resellers that reflect the Fair Information Practices, including oversight mechanisms such as the maintenance and review of audit logs detailing queries of information reseller databases—to improve accountability for agency use of such information.

AGENCY COMMENTS AND OUR EVALUATION

We received written comments on a draft of this chapter from the Justice's Assistant Attorney General for Administration (reproduced in appendix III), from the Director of the DHS Departmental GAO/OIG Liaison Office (reproduced in appendix IV), from the Commissioner of SSA (reproduced in appendix V), and from State's Assistant Secretary and Chief Financial Officer (reproduced in appendix VI). We also received comments via E - mail from staff of OMB's Office of Information and Regulatory Affairs. Justice, DHS, SSA, and OMB all generally agreed with the chapter and described actions initiated to address our recommendations. Justice and SSA also provided technical comments, which has been incorporated in the final chapter as appropriate.

In its comments, Justice agreed that revised or additional guidance and policy could be created to address unique issues presented by use of personal information obtained from resellers. However, noting that the Privacy Act allows law enforcement agencies to exempt certain records from provisions of the law that reflect aspects of the Fair Information Practices, Justice recommended that prior to issuance of any new or revised policy, careful consideration be given to the balance struck in the Privacy Act on applying the Fair Information Practices to law enforcement data. We recognize that law enforcement purposes are afforded the opportunity for exemptions from some of the provisions of the Privacy Act. The chapter acknowledges this fact. We also agree and acknowledge in the chapter that the Fair Information Practices serve as a framework of principles for balancing the need for privacy with other public policy interests, such as national security and law enforcement.

DHS also agreed on the importance of guidance to federal agencies on the use of reseller information and stated that it is working diligently on finalizing a DHS policy for such use.

The agency commented that its Privacy Office has been reviewing the use and appropriate privacy protections for reseller data, including conducting a 2-day public workshop on the subject in September 2005. DHS also noted that it had just issued departmentwide guidance on the conduct of privacy impact assessments in March 2006, which include directions relevant to the collection and use of commercial data. We have made changes to the final chapter to reflect the recent issuance of the DHS guidance.

SSA noted in its comments that it had established internal controls, including audit trails of systems usage, to ensure that information is not improperly disclosed. SSA also stated that it would amend relevant systemof-record notices to reflect use of information resellers and would explore options for enhancing its policies and internal controls over information obtained from resellers.

State interpreted our draft chapter to "rest on the premise that records from 'information resellers' should be accorded special treatment when compared with sensitive information from other sources." State indicated that it does not distinguish between types of information or sources of information in complying with privacy laws. However, our chapter does not suggest that data from resellers should receive special treatment. Instead, our chapter takes the widely accepted Fair Information Practices as a universal benchmark of privacy protections and assesses agency practices in comparison with them. State also interpreted our draft chapter to state that fraud detection, as a purpose for collecting personal information, is not related to law enforcement. However, the draft does not make such a claim. We have categorized agency uses of personal information based on descriptions provided by agencies and have categorized fraud detection uses separately from law enforcement to provide insight into different types of uses. We do not claim the two uses are unrelated. Finally, the department stated that in its view, it would be bad policy to require specification of sources such as data resellers in agency system of records notices. In contrast, we believe that adding clarity and specificity about sources is in the spirit of the purpose specification practice and note that DHS has recently issued guidance on privacy impact assessments that is consistent with this view.

OMB stated that, based on a staff-level meeting of agency privacy experts, it believes agencies recognize that when personal data are brought into their systems, this fact must be reflected in their privacy impact assessments and system-of-record notices. We do not find this observation inconsistent with our findings. We found, however, that inconsistencies occurred in agencies' determinations of when or whether reseller information was actually brought into their systems, as opposed to being merely "accessed" on an ad-hoc basis. We believe clarification of this issue is important. OMB further stated that agencies have procedures in place to verify commercial data before they are used in decisions involving the granting or recoupment of benefits or entitlements. Again, this is not inconsistent with the results of our review. Finally OMB stated that it would discuss its guidance with agency senior officials for privacy to determine whether additional guidance concerning reseller data is needed.

COMMENTS FROM INFORMATION RESELLERS

We also obtained comments on excerpts of our draft chapter from the five information resellers we reviewed. General comments made by resellers and our evaluation are summarized below:

- Several resellers raised concerns about our reliance on the OECD version of the Fair Information Practices as a framework for assessing their privacy policies and business practices. They suggested that it would be unreasonable to require them to comply with aspects of the Fair Information Practices that they believe were intended for other types of users of personal information, such as organizations that collect information directly from consumers. Further, they commented that our draft summary appeared to treat strict adherence to all of the Fair Information Practices as if it were a legally binding requirement. In several cases, they suggested that it would be more appropriate for us to use the privacy framework developed by the Asia-Pacific Economic Cooperation (APEC) organization in 2004, because the APEC framework is more recent and because it explicitly states that it has limited applicability to publicly available information.
- As discussed in our chapter, the OECD version of the Fair Information Practices is widely used and cited within the federal government as well as internationally. In addition, the APEC privacy framework, which was developed as a tool for encouraging the development of privacy protection in the Asia Pacific region, acknowledges that the OECD guidelines are still relevant and "in many ways represent the international consensus on what constitutes honest and trustworthy treatment of personal information."65 Further, our use of the OECD guidelines is as an analytical framework for identifying potential privacy issues for further consideration by Congress—not as legalistic compliance criteria. The chapter states that the Fair Information Practices are not precise legal requirements; rather they provide a framework of principles for balancing the needs for privacy against other public policy interests, such as national security, law enforcement, and administrative efficiency. In conducting our analysis, we noted that the nature of the reseller business is largely at odds with the principles of collection limitation, data quality, purpose specification, and use limitation. We also noted that resellers are not currently required to follow the Fair Information Practices and that for resellers to more fully embrace them could require that they change the way they do business. We recognize that it is important to achieve an appropriate balance between the benefits of resellers' services and the public's right to privacy and point out that, as Congress weighs various legislative options, it will be critical to determine an appropriate balance. We have made changes in this chapter to clarify that we did not attempt to make determinations of whether or how information reseller practices should change and that such determinations are a matter of policy based on balancing the public's right to privacy with the value of reseller services.
- Several information resellers stated that the draft did not take into account that public record information is freely available. For example, one reseller stated that public records should be understood by consumers to be open to all for any use not

prohibited by state or federal law. Another stated that information resellers merely effectuate the determination made by governmental entities that public records should be open to all.

However, the views expressed by the resellers do not take into account several important factors. First, resellers collect information for their products from a variety of sources, including information provided by consumers to businesses. Resellers products are not based exclusively on public records. Thus a consideration of protections for public record information does not take the place of a full assessment of the information reseller business. Second, resellers do not merely pass on public record information as they find it; they aggregate information from many different sources to create new information products, and they make the information much more readily available than it would be if it remained only in paper records on deposit in government facilities. The aggregation and increased accessibility provided by resellers raises privacy concerns that may not apply to the original paper-based public records. Finally, it is not clear that individuals give up all privacy rights to personal information contained in public records. The Supreme Court has expressed the opinion in the past that individuals retain a privacy interest in publicly released personal information. We therefore believe it is important to assess the status of privacy protections for all personal information being offered commercially to the government so that informed policy decisions may be made about the appropriate balance between resellers' services and the public's right to privacy.

- Several resellers also noted that the draft chapter did not address the complexity of the reseller business—the extent to which resellers' businesses vary among themselves and overlap with consumer reporting agencies. We have added text addressing this in the final chapter.

The resellers also provided technical comments, which were incorporated in the final chapter as appropriate.

LIST OF REQUESTERS

The Honorable F. James Sensenbrenner, Jr.
Chairman

The Honorable John Conyers, Jr.
Ranking Minority Member
Committee on the Judiciary
House of Representatives

The Honorable Steve Chabot
Chairman

The Honorable Jerrold Nadler
Ranking Minority Member
Subcommittee on the Constitution Committee on the Judiciary
House of Representatives

The Honorable Bill Nelson
Ranking Minority Member Subcommittee on International Operations and Terrorism,
Committee on Foreign Relations United States Senate

The Honorable Bennie G. Thompson
Ranking Minority Member
Committee on Homeland Security House of Representatives

The Honorable Zoe Lofgren
Ranking Minority Member
Subcommittee on Intelligence, Information Sharing, and Terrorism Risk Assessment
Committee on Homeland Security
House of Representatives

The Honorable Loretta Sanchez
Ranking Minority Member
Subcommittee on Economic Security, Infrastructure Protection, and Cybersecurity
Committee on Homeland Security
House of Representatives

REFERENCES

[1]   For purposes of this chapter, the term personal information encompasses all information associated with an individual, including both identifying and nonidentifying information. Personally identifying information, which can be used to locate or identify an individual, includes such things as names, aliases, and agency-assigned case numbers. Nonidentifying personal information includes such things as age, education, finances, criminal history, physical attributes, and gender.

[2]   The Privacy Act of 1974, Pub. L. No. 93-579, 88 Stat. 1896 (codified as amended at 5 U.S.C. § 552a) provides safeguards against an invasion of privacy through the misuse of records by federal agencies and allows citizens to learn how their personal information is collected, maintained, used, and disseminated by the federal government.

[3]   Congress used the committee's final report as a basis for crafting the Privacy Act of 1974. See Records, Computers and the Rights of Citizens: Report of the Secretary's Advisory Committee on Automated Personal Data Systems (Washington, D.C.: U.S. Department of Health, Education, and Welfare, July 1973).

[4]   Descriptions of these principles are shown in table 2.

[5]   The five information resellers we reviewed were ChoicePoint, LexisNexis, Acxiom, Dun and Bradstreet, and West. While these resellers were all reported by federal

agencies to be sources of personal information, their businesses vary. A discussion of this variance in business practices appears in the background section of this chapter. Our results may not apply to other resellers who do very little or no business with these federal agencies.

[6]   ChoicePoint, LexisNexis, and Acxiom.

[7]   This figure may include uses that do not involve personal information. Except for instances where the reported use was primarily for legal research, agency officials were unable to separate the dollar values associated with use of personal information from uses for other purposes (e.g., LexisNexis and West provide news and legal research in addition to public records).

[8]   In certain circumstances, laws restrict the collection and use of specific kinds of personal information. For example, the Fair Credit Reporting Act regulates access to and use of consumer information under certain circumstances.

[9]   Credit header data are the nonfinancial identifying information located at the top of a credit report, such as name, current and prior addresses, telephone number, and Social Security number.

[10] Under the Privacy Act of 1974, the term "routine use" means (with respect to the disclosure of a record) the use of such a record for a purpose that is compatible with the purpose for which it was collected. 5 U.S.C. § 552a (a(7)).

[11] 5 U.S.C. § 552a(m).

[12] OMB, OMB Guidance for Implementing the Privacy Provisions of the E-Government Act of 2002, M-03-22 (Sept. 26, 2003).

[13] FISMA, Title III, E-Government Act of 2002, Pub. L. No. 107-347 (Dec. 17, 2002).

[14] OMB, "Privacy Act Implementation: Guidelines and Responsibilities," Federal Register, Volume 40, Number 132, Part III, pages 28948-28978 (Washington, D.C.: July 9, 1975). Since the initial Privacy Act guidance of 1975, OMB periodically has published additional guidance. Further information regarding OMB Privacy Act guidance can be found on the OMB Web site at http://www.whitehouse.gov/omb/inforeg/infopoltech.html.

[15] States that enacted breach of information legislation in 2005 include Arkansas, Connecticut, Delaware, Florida, Georgia, Illinois, Indiana (applies to state agencies only), Louisiana, Maine, Minnesota, Montana, Nevada, New Jersey, New York, North Carolina, North Dakota, Ohio, Pennsylvania, Rhode Island, Tennessee, Texas, and Washington.

[16] Records, Computers and the Rights of Citizens: Report of the Secretary's Advisory Committee on Automated Personal Data Systems, (Washington, D.C.: U.S. Department of Health, Education, and Welfare, July 1973).

[17] OECD, Guidelines on the Protection of Privacy and Transborder Flow of Personal Data (Sept. 23, 1980). The OECD plays a prominent role in fostering good governance in the public service and in corporate activity among its 30 member countries. It produces internationally agreed-upon instruments, decisions, and recommendations to promote rules in areas where multilateral agreement is necessary for individual countries to make progress in the global economy.

[18] European Union Data Protection Directive ("Directive 95/46/EC of the European Parliament and of the Council of 24 October 1995 on the Protection of Individuals with

Regard to the Processing of Personal Data and the Free Movement of Such Data") (1995).

[19] "Report on OECD Guidelines Program," Memorandum from Bernard Wunder, Jr., Assistant Secretary for Communications and Information, Department of Commerce (Oct. 30, 1981).

[20] Privacy Office Mission Statement, U.S. Department of Homeland Security; "Privacy Policy Development Guide," Global Information Sharing Initiative, U.S. Department of Justice, www.it.ojp.gov/global (Sept. 2005); "Homeless Management Information Systems, U.S. Department of Housing and Urban Development (Federal Register, July 30, 2004); and "Options for Promoting Privacy on the National Information Infrastructure," Health and Human Services Privacy Committee, Office of the Assistant Secretary for Planning and Evaluation, Department of Health and Human Services (April 1997).

[21] The Federal Enterprise Architecture is intended to provide a common frame of reference or taxonomy for agencies' individual enterprise architecture efforts and their planned and ongoing information technology investment activities. An enterprise architecture is a blueprint, defined largely by interrelated models, that describes (in both business and technology terms) an entity's "as is" or current environment, its "to be" or future environment, and its investment plan for transitioning from the current to the future environment.

[22] H.R. 4127; introduced by Representative Clifford B. Stearns on October 25, 2005.

[23] S. 1789; introduced by Senator Arlen Specter on September 29, 2005, and reported from the Senate Judiciary Committee on November 17, 2005.

[24] This figure comprises contracts and task orders with information resellers that included the acquisition and use of personal information. However, some of these funds may have been spent on uses that do not involve personal information; we could not omit all such uses because agency officials were not always able to separate the amounts associated with use of personal information from those for other uses (e.g., LexisNexis and West provide news and legal research in addition to public records). In some instances, where the reported use was primarily for legal research, we omitted these funds from the total.

[25] GSA's Federal Supply Schedule allows agencies to take advantage of prenegotiated contracts with a variety of vendors, including information resellers.

[26] A GSA schedule blanket purchase agreement simplifies the filling of recurring needs for supplies or services, while leveraging a customer's buying power by taking advantage of quantity discounts, saving administrative time, and reducing paperwork.

[27] The ChoicePoint blanket purchase agreement is also available to non-Justice agencies, whose use accounted for approximately $2.8 million in fiscal year 2005.

[28] The total value of ChoicePoint, LexisNexis, and West contracts—$24.7 million—exceeds the value of $19 million reported above because this figure omits the $2.8 million used by non-Justice agencies (see footnote 27) as well as uses that were reported not to involve personal information. Justice officials responsible for administering the departmentwide contracts with LexisNexis and West reported that these agreements are used by multiple components whose business needs vary and may not require use of databases that include public records about individuals. In cases

where Justice officials were able to separate these costs, we omitted these costs from the total.

[29] GAO, Data Mining: Agencies Have Taken Key Steps to Protect Privacy in Selected Efforts, but Significant Compliance Issues Remain, GAO-05-866 (Washington, D.C.: Aug. 15, 2005).

[30] The U.S. Marshals Service is the federal government's primary agency for conducting investigations involving escaped federal prisoners; probation, parole, and bond violators; and fugitives named in warrants generated during drug investigations.

[31] DEA's mission involves enforcing laws pertaining to the manufacture, distribution, and dispensing of legally produced controlled substances.

[32] The personal information contained in this information reseller database is limited to the prescribing doctor and does not contain personal patient information.

[33] To ensure that criminals do not benefit financially from their illegal acts, federal law provides that profits from drug-related crimes, as well as property used to facilitate certain crimes, are subject to forfeiture to the government.

[34] Justice for All Act of 2004, Pub. L. No. 108-405 (Oct. 30, 2004). Section 102 of the act establishes rights for crime victims including the right to "reasonable, accurate, and timely notice of any public court proceeding, or any parole proceeding, involving the crime of or any release or escape of the accused."

[35] For an assessment of privacy issues associated with the Secure Flight commercial data test, see GAO, Aviation Security: Transportation Security Administration Did Not Fully Disclose Uses of Personal Information during Secure Flight Program Testing in Initial Privacy Notices, but Has Recently Taken Steps to More Fully Inform the Public, GAO-05- 864R (Washington, D.C.: July 22, 2005).

[36] Skiptracing is the process of locating people who have fled in order to avoid paying debts.

[37] Although the Library of Congress indicated that the Department of State also used FEDLINK contracts with Dun and Bradstreet and LexisNexis, State officials reported that their use of these contracts did not involve access to personal information.

[38] Section 103 of Pub. L. 106-481 (2 U.S.C. 182c) establishes FEDLINK as a revolving fund. The law authorizes the FEDLINK revolving fund to provide "the procurement of commercial information services, publications in any format, and library support services, related accounting services, related education, information and support services" to federal offices and to other organizations entitled to use federal sources of supply.

[39] We reviewed the practices of five major information resellers: ChoicePoint, LexisNexis, Acxiom, Dun and Bradstreet, and West. While these resellers were all reported by federal agencies to be sources of personal information, their businesses vary. A discussion of this variance in business practices appears in the background section of this report.

[40] Resellers are constrained from collecting certain types of information and aggregating it with other personal information. For example, the Fair Credit Reporting Act and the Gramm-Leach-Bliley Act constrain the collection and use of personal information, such as financial information.

[41] Several information resellers reported that if the inaccuracy was a result of their error (e.g., transposing numbers or letters or incorrectly aggregating information), they would correct the data in their databases.

[42] One reseller reported that it maintains discrete databases developed and tailored toward its specific product offerings in marketing, fraud prevention, and directory services. These product offerings are geared toward specific clients. For example, the reseller's fraud prevention product makes use of public record and publicly available information as well as credit header information. The fraud prevention product provides identity verification and investigative tools primarily to the financial and insurance industries and to law enforcement agencies involved in fraud or criminal investigations. Within the four agencies, use of this reseller was reported only as part of TSA's Secure Flight commercial data test.

[43] While a significant amount of reseller information comes from public records, resellers also use private companies, including other companies that aggregate information, as suppliers. For example, a reseller may contract with another private firm to obtain telephone book information. Further, resellers may contract with other private firms to collect information from public records sources.

[44] In its settlement with ChoicePoint, the Federal Trade Commission alleged violations of the Fair Credit Reporting Act and section 5 of the Federal Trade Commission Act. Section 5 of the act prohibits "unfair or deceptive acts or practices in or affecting commerce." The Federal Trade Commission can issue orders, obtain injunctions, impose civil penalties, and undertake civil actions to enforce the act. 5 U.S.C. § 45.

[45] One reseller reported that, for certain products, it will delete information that has been identified as inaccurate. For example, if the reseller is able to verify that data contained within its directory or fraud products are inaccurate, it will delete the inaccurate data and keep a record of this in a maintenance file so the erroneous data are not reentered at a future date.

[46] 5 U.S.C. § 552a (e)(1). The Privacy Act (at § 552a (j) and (k)) allows agencies to claim an exemption from this provision if the records are used for certain purposes. For example, records compiled for criminal law enforcement purposes or for a broader category of investigative records compiled for criminal or civil law enforcement purposes can be exempted from this requirement.

[47] In two cases, agency components used reseller data to conduct broader searches for previously unidentified criminal behavior. These two cases were an application at DEA used to identify potential prescription drug fraud and efforts by Citizenship and Immigration Services to detect large patterns of potential fraud through address searches and other queries.

[48] 5 U.S.C. § 552a(e)(5). The Privacy Act allows agencies to claim an exemption from this provision of the act for certain designated purposes. For example, records compiled for criminal law enforcement purposes can be exempt from this provision. A broader category of investigative records compiled for criminal or civil law enforcement purposes cannot be exempt from this provision.

[49] Such uses are referred to as "routine uses" in the Privacy Act, 5 U.S.C. § 552a (a(7)) and (b).

[50] The task force's partner agencies include ICE, the Department of Defense Counterintelligence Field Activity Office, the Office of Personnel Management, and members of the intelligence community.

[51] 5 U.S.C. § 552a(e)(10).

[52] Although we did not assess the effectiveness of information security or compliance with FISMA at any agency as part of this review, we have previously reported on weaknesses in almost all areas of information security controls at 24 major agencies, including Justice, DHS, State, and SSA. For additional information see GAO, Information Security: Weaknesses Persist at Federal Agencies Despite Progress Made in Implementing Related Statutory Requirements, GAO-05-552 (Washington, D.C.: July 15, 2005) and Information Security: Department of Homeland Security Needs to Fully Implement Its Security Program, GAO-05-700 (Washington, D.C.: June 17, 2005).

[53] 5 U.S.C. § 552a(e)(4)(C) and (I). The Privacy Act allows agencies to claim an exemption from identifying the categories of sources of records for records compiled for criminal law enforcement purposes, as well as for a broader category of investigative records compiled for criminal or civil law enforcement purposes.

[54] The act provides for its requirements to apply to government contractors when agencies contract for the operation by or on behalf of the agency, a system of records to accomplish an agency function. 5 U.S.C. § 552a(m).

[55] As we previously reported, this notice did not fully disclose the scope of the use of reseller data during the test phase. See GAO-05-864R.

[56] The Privacy Act allows agencies to claim an exemption from identifying the categories of sources of records for records compiled for criminal law enforcement purposes as well as for a broader category of investigative records compiled for criminal or civil law enforcement purposes. 5 U.S.C. § 552a (j) and (k). One system of records notice for the Treasury Enforcement Communications System (the system identified by ATF as covering their investigative case files) claimed such an exemption. The Department of State identifies categories of sources in the system of records notices it identified but does not specifically identify use of reseller data. The State system of records notices also claim an exemption from identifying categories of sources but invoke that exemption only under certain circumstances (e.g., to the extent that a specific investigation would be compromised).

[57] The notice was last updated in October 2002, before the service and benefit functions of the U.S. Immigration and Naturalization Service transitioned into DHS as U.S. Citizenship and Immigration Services.

[58] The Privacy Act allows agencies to claim exemptions if the records are used for certain purposes. 5 U.S.C. § 552a (j) and (k). For example, records compiled for criminal law enforcement purposes can be exempt from the access and correction provisions. In general, the exemptions for law enforcement purposes are intended to prevent the disclosure of information collected as part of an ongoing investigation that could impair the investigation or allow those under investigation to change their behavior or take other actions to escape prosecution.

[59] The E-Government Act requires agencies, if practicable, to make privacy impact assessments publicly available through agency Web sites, publication in the Federal Register, or by other means. Pub. L. No. 107-347, § 208 (b)(1)(B)(iii).

[60] The agency components that identified preparation of PIAs for systems or programs making use of information reseller data included USCIS for its Fraud Tracking System, TSA for its Secure Flight commercial data test, and FBI's FTTTF, which reported that it was in the process of finalizing a PIA. Only the PIA for TSA's test specifically identified the use of commercial data. We were unable to determine if FTTTF's PIA identified the use of commercial data since it was not yet final.

[61] OMB, Guidance for Implementing the Privacy Provisions of the E-Government Act of 2002, Memorandum M-03-22 (Washington, D.C.: Sept. 26, 2003).

[62] The DHS Privacy Officer position was created by the Homeland Security Act of 2002, Pub. L. No 107-296, § 222, 116 Stat. 2155. The Privacy Officer is responsible for, among other things, "assuring that the use of technologies sustain[s], and do[es] not erode privacy protections relating to the use, collection, and disclosure of personal information, and assuring that personal information contained in Privacy Act systems of records is handled in full compliance with Fair Information Practices as set out in the Privacy Act of 1974."

[63] Department of Homeland Security Privacy Office, Privacy Impact Assessments: Official Guidance (March 2006), p. 34.

[64] USCIS officials stated that the PIA for the Fraud Tracking System, now called the Fraud Detection and National Security System, would be updated on an incremental basis and that a future update would identify information resellers as a data source.

[65] Asia-Pacific Economic Cooperation, APEC Privacy Framework, Version 4 (Santiago, Chile: Nov. 17-18, 2004), p. 4.

[66] Dun and Bradstreet specializes in business information, which may contain personal information on business owners.

[67] We obtained information on policies and practices from the following major components of Justice and DHS. For Justice: Bureau of Alcohol Tobacco, Firearms, and Explosives, Drug Enforcement Administration, Executive Office for U.S. Attorneys, Executive Office of the U.S. Trustees, Federal Bureau of Investigation, and the U.S. Marshals Service. For DHS: U.S. Citizenship and Immigration Services, U.S. Immigration and Customs Enforcement, Transportation Security Administration, U.S. Secret Service, U.S. Customs and Border Protection, and the Federal Emergency Management Agency. We did not obtain information on policies and management practices for smaller components.

APPENDIX I

Objectives, Scope, and Methodology

Our objectives were to determine the following:

- how the Departments of Justice, Homeland Security, and State and the Social Security Administration are making use of personal information obtained through contracts with information resellers;

- the extent to which the information resellers providing personal information to these agencies have policies and practices in place that reflect widely accepted principles for protecting the privacy and security of personal information; and
- the extent to which these agencies have policies and practices in place for handling information reseller data that reflect widely accepted principles for protecting the privacy and security of personal information.

To address our objectives, we identified and reviewed applicable laws such as the Privacy Act of 1974 and the E-Government Act, agency policies and practices, and the widely accepted privacy principles embodied in the Organization for Economic Cooperation and Development (OECD) version of the Fair Information Practices. Working with liaisons at the four federal agencies we were requested to review, we identified officials responsible for the acquisition and use of personal information from information resellers. Through these officials, we obtained applicable contractual documentation such as statements of work, task orders, blanket purchase agreements, purchase orders, interagency agreements, and contract terms and conditions.

To address our first objective, we obtained and reviewed contract vehicles covering federal agency use of information reseller services for fiscal year 2005. We also reviewed applicable General Services Administration (GSA) schedule and Library of Congress FEDLINK contracts with information resellers that agencies made use of by various means, including through issuance of blanket purchase agreements, task orders, purchase orders, or interagency agreements. We analyzed the contractual documentation provided to determine the nature, scope, and dollar amounts associated with these uses, as well as mechanisms for acquiring personal information. In an effort to identify all relevant instances of agency use of information resellers and related contractual documents, we developed a list of structured questions to address available contract documents, uses of personal information, and applicable agency guidance. We provided these questions to agency officials and held discussions with them to help ensure that they provided all relevant information on uses of personal information from information resellers. To further ensure that relevant contract vehicles were identified, we asked major information resellers about their business with the four agencies. We also interviewed officials from GSA and the Library of Congress to discuss the mechanisms available to federal agencies for acquiring personal information and to identify any additional uses of these mechanisms by the four agencies.

To further address our first objective, we categorized agency use of information resellers into five categories: counterterrorism, debt collection, fraud detection/prevention, law enforcement, and other. These categorizations were based on the component and applicable program's mission, as well as the specific reported use of the contract. In identifying relevant uses of information resellers, we were unable to identify small purchases (e.g., purchases below $2,500), as agencies do not track this information centrally. In addition, to the extent practicable, we excluded uses that generally did not involve the use of personal information. For example, officials from several component agencies reported that their use of the LexisNexis and West services was primarily for legal research rather than for public records information. In other cases, reported amounts may reflect uses that do not involve personal information because agencies were unable to separate such uses from uses involving personal information.

To address our second objective, we obtained and reviewed relevant private sector laws and guidance, such as the Gramm-Leach-Bliley Act, the Fair Credit Reporting Act, and the Fair Information Practices. We also identified major information resellers in agency contractual agreements for personal information and held interviews with officials from these companies, including Acxiom, ChoicePoint, Dun and Bradstreet, [66] LexisNexis, and West, to discuss security, quality controls, and privacy policies. In addition, we conducted site visits at Acxiom, ChoicePoint, and LexisNexis, and obtained written responses to related questions from West. These five resellers accounted for approximately 95 percent of the dollar value of all reported contracts with resellers. To determine the extent that they reflect widely accepted Fair Information Practices, we reviewed and compared information reseller's privacy policies and procedures with these principles. In conducting our analysis, we identified the extent to which reseller practices were consistent with the key privacy principles of the Fair Information Practices. We also assessed the effect of any inconsistencies; however, we did not attempt to make determinations of whether or how information reseller practices should change. Such determinations are a matter of policy based on balancing the public's right to privacy with the value of services provided by resellers to customers such as government agencies.

To address our third objective, we identified applicable guidelines and management controls regarding the acquisition, maintenance, and use of personal information from information resellers at each of the four agencies. We also interviewed agency officials, including acquisition and program staff, to further identify relevant policies and procedures. Our assessment of overall agency application of the Fair Information Practices was based on the policies and procedures of major components at each of the four agencies.[67] We also conducted interviews at the four agencies with senior agency officials designated for privacy as well as officials of the Office of Management and Budget (OMB) to obtain their views on the applicability of federal privacy laws (including the Privacy Act of 1974 and the E-Government Act of 2002) and related guidance on agency use of information resellers. In addition, we compared relevant policies and management practices with the Fair Information Practices.

We assessed the overall application of the principles of the Fair Information Practices by agencies according to the following categories:

1. *General.* We assessed the application as general if the agency had policies or procedures to address all major aspects of a particular principle.
2. *Uneven.* We assessed the application as uneven if the agency had policies or procedures that addressed some but not all aspects of a particular principle or if some but not all components and agencies had policies or practices in place addressing the principle.

We performed our work at the Departments of Homeland Security, Justice, and State in Washington, D.C.; at the Social Security Administration in Baltimore, Maryland; Acxiom Corporation in Little Rock, Arkansas; ChoicePoint in Alpharetta, Georgia; Dun and Bradstreet in Washington, D.C.; and LexisNexis in Washington, D.C., and Miamisburg, Ohio. Our work was conducted from May 2005 to March 2006 in accordance with generally accepted government auditing standards.

APPENDIX II

Federal Laws Affecting Information Resellers

Major laws that affect information resellers include the Gramm-LeachBliley Act, the Drivers Privacy Protection Act, the Health Insurance Portability and Accountability Act, the Fair Credit Reporting Act, and the Fair and Accurate Credit Transactions Act. Their major privacy related provisions are briefly summarized below.

Gramm-Leach-Bliley Act

The Gramm-Leach-Bliley Act requires financial institutions (e.g., banks, insurance, and investment companies) to give consumers privacy notices that explain the institutions' information-sharing practices (P.L. 106-102 (1999), Title V, 15 U.S.C. 6801). In turn, consumers have the right to limit some, but not all, sharing of their nonpublic personal information. Financial institutions are permitted to disclose consumers' nonpublic personal information without offering them an opt-out right in a number of circumstances including the following:

- to effect a transaction requested by the consumer in connection with a financial product or service requested by the consumer; maintaining or servicing the consumer's account with the financial institution or another entity as part of a private label credit card program or other extension of credit; or a securitization, secondary market sale, or similar transaction;
- with the consent or at the direction of the consumer;
- to protect the confidentiality or security of the consumer's records; to prevent fraud; for required institutional risk control or for resolving customer disputes or inquiries; to persons holding a legal or beneficial interest relating to the consumer; or to the consumer's fiduciary;
- to provide information to insurance rate advisory organizations, guaranty funds or agencies, rating agencies, industry standards agencies, and the institution's attorneys, accountants, and auditors;
- to the extent specifically permitted or required under other provisions of law and in accordance with the Right to Financial Privacy Act of 1978, to law enforcement agencies, self-regulatory organizations, or for an investigation on a matter related to public safety;
- to a consumer reporting agency in accordance with the Fair Credit Reporting Act or from a consumer report reported by a consumer reporting agency;
- in connection with a proposed or actual sale, merger, transfer, or exchange of all or a portion of a business if the disclosure concerns solely consumers of such business; and
- to comply with federal, state, or local laws; an investigation or subpoena; or to respond to judicial process or government regulatory authorities.

Driver's Privacy Protection Act

The Driver's Privacy Protection Act generally prohibits the disclosure of personal information by state departments of motor vehicles. (P.L. 103-322 (1994), 18 U.S.C. § 272 1-2725). It also specifies a list of exceptions when personal information contained in a state motor vehicle record may be disclosed. These permissible uses include the following:

- for use by any government agency in carrying out its functions;
- for use in connection with matters of motor vehicle or driver safety and theft; motor vehicle emissions; motor vehicle product alterations, recalls, or advisories; motor vehicle market research activities;
- for use in the normal course of business by a legitimate business, but only to verify the accuracy of personal information submitted by the individual to the business and, if such information is not correct, to obtain the correct information but only for purposes of preventing fraud by pursuing legal remedies against, or recovering on a debt or security interest against, the individual;
- for use in connection with any civil, criminal, administrative, or arbitral proceeding in any federal, state, or local court or agency;
- for use in research activities;
- for use by any insurer or insurance support organization in connection with claims investigation activities;
- for use in providing notice to the owners of towed or impounded vehicles;
- for use by a licensed private investigative agency for any purpose permitted under the act;
- for use by an employer or its agent or insurer to obtain information relating to the holder of a commercial driver's license;
- for use in connection with the operation of private toll transportation facilities;
- for any other use, if the state has obtained the express consent of the person to whom a request for personal information pertains;
- for bulk distribution of surveys, marketing, or solicitations, if the state has obtained the express consent of the person to whom such personal information pertains;
- for use by any requester, if the requester demonstrates that it has obtained the written consent of the individual to whom the information pertains; and
- for any other use specifically authorized under a state law, if such use is related to the operation of a motor vehicle or public safety.

Health Insurance Portability and Accountability Act

The Health Insurance Portability and Accountability Act of 1996 (P.L. 104- 191) made a number of changes to laws relating to health insurance. It also directed the Department of Health and Human Services to issue regulations to protect the privacy and security of personally identifiable health information. The resulting privacy rule (45 C.F.R. Part 164) defines certain rights and obligations for covered entities (e.g., health plans and health care providers) and individuals, including the following:

- giving individuals the right to be notified of privacy practices and to inspect, copy, request correction, and have an accounting of disclosures of health records, except for specified exceptions;
- setting limits on the use of health information apart from treatment, payment, and health care operations (e.g., for marketing) without the individual's authorization;
- permitting disclosure of health information without the individual's authorization for purposes of public health protection; health oversight; law enforcement; judicial and administrative proceedings; approved research activities; coroners, medical examiners, and funeral directors; workers' compensation programs, government abuse, neglect, and domestic violence authorities; organ transplant organizations; government agencies with specified functions, e.g., national security activities; and as required by law;
- requiring that authorization forms contain specific types of information, such as a description of the health information to be used or disclosed, the purpose of the use or disclosure, and the identity of the recipient of the information; and
- requiring covered entities to take steps to limit the use or disclosure of health information to the minimum necessary to accomplish the intended purpose, unless authorized or under certain circumstances.

Fair Credit Reporting Act

The Fair Credit Reporting Act (P.L. 91-508, 1970, 15 U.S.C. § 1681) governs the use of personal information by consumer reporting agencies, which are individuals or entities that regularly assemble or evaluate information about individuals for the purpose of furnishing consumer reports to third parties. The act defines a consumer report as any communication by a consumer reporting agency about an individual's credit worthiness, character, reputation, characteristics, or mode of living and permits its use only in the following situations:

- as ordered by a court or federal grand jury subpoena;
- as instructed by the consumer in writing;
- for the extension of credit as a result of an application from a consumer or the review or collection of a consumer's account;
- for employment purposes, including hiring and promotion decisions, where the consumer has given written permission;
- for the underwriting of insurance as a result of an application from a consumer;
- when there is a legitimate to determine a consumer's eligibility for a license or other benefit granted by a governmental instrumentality required by law to consider an applicant's financial responsibility or status;
- for use by a potential investor or servicer or current insurer in a valuation or assessment of the credit or prepayment risks associated with an existing credit obligation; and
- for use by state and local officials in connection with the determination of child support payments, or modifications of enforcement thereof.

The act generally limits the amount of time negative information can be included in a consumer report to no more than 7 years, or 10 years in the case of bankruptcies. Under the act, individuals have a right to access all information in their consumer reports; a right to know who obtained their report during the previous year or two, depending on the circumstances; and a right to dispute the accuracy of any information about them.

Fair and Accurate Credit Transactions Act

The Fair and Accurate Credit Transactions Act (P.L. 108-159, 2003) amended the Fair Credit Reporting Act, extending provisions to improve the accuracy of personal information assembled by consumer reporting agencies and better provide for the fair use of and consumer access to personal information. The act's provisions include the following:

- consumers may request a free annual credit report from nationwide consumer reporting agencies, to be made available no later than 15 days after the date on which the request is received;
- persons furnishing information about individuals to consumer reporting agencies, and resellers of consumer reports, must have polices and procedures for investigating and correcting inaccurate information,
- consumers are given the right to prohibit business affiliates of consumer reporting agencies from using information about them for certain marketing purposes; and
- consumer reporting agencies cannot include medical information in reports that will be used for employment, credit transactions, or insurance transactions unless the consumer consents to such disclosures.

APPENDIX III

Comments from the Department of Justice

U.S. Department of Justice

MAR 1 7 2006

Washington, D.C. 20530

Linda Koontz
Director, Information Management Issues
U.S. Government Accountability Office
441 G Street, NW
Washington, DC 20548

Dear Ms. Koontz:

Thank you for the opportunity to review the final draft of the Government Accountability Office (GAO) report entitled Privacy: Opportunities Exist for Agencies and Information Resellers to More Fully Adhere to Key Principles (GAO-06-421/310228). The draft was reviewed by 16 components of the Department of Justice (DOJ) who had participated in this review. Earlier today, the DOJ provided you technical comments to be incorporated in the report as appropriate. This letter constitutes the formal comments of the DOJ, and I request that it be included in the final report.

The DOJ is committed to protecting the privacy rights of individuals in the course of its counterterrorism and law enforcement mission. To spearhead this effort, the DOJ has recently appointed a Chief Privacy and Civil Liberties Officer (CPCLO) to oversee and administer the DOJ's privacy functions. The DOJ is also establishing a departmental Privacy and Civil Liberties Board to assist the CPCLO in ensuring that the DOJ's activities are carried out in a way that continues to fully protect the privacy and civil liberties of all Americans.

As the GAO report points out, the recent security breaches involving information resellers have highlighted the public's concerns regarding personal data maintained by such resellers and led to the GAO's review of the use of personal information from information resellers by the DOJ, as well as the DOJ's policies and practices for handling such information. The DOJ recognizes the unique issues presented by reseller information and agrees that additional measures could be taken regarding its use, in the form of revised or additional guidance and policy. At the same time, the DOJ also recognizes the need to consider agency resources, competing mission priorities, and the privacy protections that are already in place as a result of the DOJ's compliance with the Privacy Act of 1974, 5 U.S.C. §552a.

Ms. Linda Koontz   2

In recognition of the variety of government operations (such as law enforcement and intelligence), the Privacy Act incorporated some, but not all, of the Fair Information Practices.[1] Law enforcement may use the regulatory process to exempt certain records from some of the requirements of the Privacy Act. For example, pursuant to regulations, criminal law enforcement records may be exempted from the Privacy Act's requirement that an agency make reasonable efforts to assure that a record is accurate, complete, timely, and relevant for agency purposes, prior to disseminating that record to someone other than an agency or pursuant to FOIA. Instead of focusing on satisfying the Fair Information Practices, the more appropriate metric should be whether an agency has met the requirements of the Privacy Act.

Thus, the DOJ recommends that prior to the issuance of any new guidance or policy, a careful analysis and assessment of the degree of need for any new guidance should be conducted. That assessment should be used to ensure that the guidance is tailored in such a way as to avoid any negative impact on the DOJ's resources and competing mission priorities. Further, any new guidance or policy should be crafted in such a way as to avoid any increase in litigation risk, and to fully recognize and take into account the balance that Congress has already struck in the Privacy Act in applying Fair Information Practices to law enforcement data.

The DOJ stands willing to assist in the development of any new guidance or policy considered as a result of this effort. We look forward to working with OMB and other agencies toward a solution that strikes the proper balance between the furtherance of the DOJ's mission and the protection of individuals' privacy.

Again, we appreciate the opportunity to comment on this report. If you have any questions regarding our comments, please contact Richard Theis, Assistant Director, Audit Liaison Group, Management and Planning Staff. If you would like to discuss or receive a briefing, please contact me at (202) 514-3101.

Sincerely,

Paul R. Corts
Assistant Attorney General
  for Administration

---

[1] First proposed in 1973 by a U.S. governmental advisory committee and widely accepted as including: collection limitation, data quality, purpose specification, use limitation, security safeguards, openness, individual participation, and accountability.

APPENDIX IV

Comments from the Department of Homeland Security

U.S. Department of Homeland Security
Washington, DC 20528

Homeland
Security

March 17, 2006

Ms. Linda Koontz
Director, Information Management
Government Accountability Office
Washington, DC 20548

Dear Ms. Koontz:

Re: Draft Report GAO-06-421, Privacy: Opportunities Exist for Agencies and
Information Resellers to More Fully Adhere to Key Principles.

Thank you for the opportunity to review the draft report. The Department of Homeland
Security (DHS) and the Privacy Office commend the GAO for undertaking this important
and informative review. Certainly guidance on the collection and use of commercial
data is important for federal agencies, such as DHS. Early on in the establishment of the
DHS Privacy Office, the Department determined that one of the top three issues that
needed to be addressed was the use of private sector information for homeland security
purposes. It is an increasingly important issue, as the report notes.

To that end, the Privacy Office at DHS began its review of commercial data use and
appropriate privacy safeguards through internal DHS study and by doing outreach
publicly and in cooperation with DHS offices and other federal and private sector
partners. The Privacy Office hosted a two-day public workshop, September 8-9, 2005,
on Privacy and Technology: Government Use of Commercial Data for Homeland
Security. The agenda and full transcripts of the conference, including a review of the
application of the Privacy Act and Fair Information Practice Principles, is posted at our
website at www.dhs.gov/privacy and is available to the public and government agencies
for review. Mention of this in the final GAO report could assist the dialogue and enable
decision makers to review information and suggestions raised for appropriate use of
commercial data and challenges experienced by federal agencies.

The Department appreciates the thoughtful work of GAO in addressing current use and
practices at DHS. We would like to report that in early March 2006, and since the last
contact with GAO, updated Privacy Impact Assessment Guidance, which includes
directions relevant to the collection and use of commercial data, has been published by
the Privacy Office and distributed throughout the Department. It also is posted on both
the Department's internal and external websites. Please see *Privacy Impact Assessments,
Official Guidance 2006, Privacy Office, U.S. Department of Homeland Security.* We
respectfully suggest the GAO report could be updated to reflect this. Prior to this, the

Department did have guidance on Privacy Impact Assessments that had been distributed
in draft form in July 2005, both internally in DHS and externally with all of our federal
partners. The Department of Justice advised DHS of their intention to adopt the DHS
published guidance of March 2006.

The Department believes that our guidance, which includes questions that address the use
of commercial data, is unique in the government in this regard. As a result, we believe
the DHS Privacy Office should be given recognition in the GAO report for its efforts to
encourage transparency regarding the use of commercial data. The Department continues
to work diligently on finalizing a policy for DHS use of commercial data and expects to
have that policy in circulation shortly. The Department will continue to address the need
for transparency about the use of commercial data as part of the overall effort to
reorganize and review legacy Privacy Act systems.

We thank you again for the opportunity to review this most important report and provide
comments.

Sincerely,

Steven J. Pecinovsky
Director
Departmental GAO/OIG Liaison Office

APPENDIX V

Comments from the Social Security Administration

SOCIAL SECURITY
The Commissioner
March 17, 2006

Ms. Linda Koontz
Director, Information Management Issues
U.S. Government Accountability Office
Room 4-T-21
441 G Street, NW
Washington, D.C. 20548

Dear Ms. Koontz:

Thank you for the opportunity to review the draft report, "Privacy: Opportunities Exist For Agencies and Information Resellers to More Fully Adhere to Key Principles" (GAO-06-421). Our comments are enclosed.

If you have any questions, please have your staff contact Candace Skurnik, Director, Audit Management and Liaison Staff, at (410) 965-0374.

Sincerely,

JoAnne B. Barnhart

Enclosure

SOCIAL SECURITY ADMINISTRATION   BALTIMORE MD 21235-0001

COMMENTS OF THE SOCIAL SECURITY ADMINISTRATION (SSA) ON THE GOVERNMENT ACCOUNTABILITY OFFICE'S (GAO) DRAFT REPORT, "PRIVACY: OPPORTUNITIES EXIST FOR AGENCIES AND INFORMATION RESELLERS TO MORE FULLY ADHERE TO KEY PRINCIPLES" (GAO-06-421)

General Comments

Thank you for the opportunity to review and provide comments on this GAO draft report. We share GAO's concerns about the potential for security breaches involving information resellers and support GAO's suggestion for congressional consideration and recommendations for Executive Branch action in support of ensuring adherence to applicable laws and the Fair Information Practices relating to privacy protection.

SSA is committed to protecting privacy with regard to information the Agency maintains, including information obtained from information resellers. We have established internal controls, including audit trails of any systems usage, to ensure that any information disclosed is for proper use. In order to identify any internal control weaknesses and potential problems that could result in waste, fraud and abuse, and to ensure compliance with the Federal Managers Financial Integrity Act of 1982, SSA components regularly perform Management Control Systems Reviews mandated by SSA and the Office of Management and Budget.

GAO Recommendation

We recommend that the Attorney General, the Secretary of Homeland Security, the Secretary of State, and the Commissioner of SSA develop specific policies for the collection, maintenance, and use of personal information obtained from resellers that reflect the Fair Information Practices, including oversight mechanisms such as the maintenance and review of audit logs detailing queries of information reseller databases, to improve accountability for agency use of such information.

SSA Comment

We agree. To better address the Fair Information Practices concerning information SSA obtains from information resellers, we will amend our relevant Privacy Act systems of records notices to reflect the use of information resellers/commercial data sources.

We will also explore options for enhancing our policies and internal controls over information SSA obtains from information resellers, including options for improved audit trail maintenance and review.

APPENDIX VI

Comments from the Department of State

United States Department of State

Assistant Secretary and Chief Financial Officer

Washington, D.C. 20520

MAR 20 2006

Ms. Jacquelyn Williams-Bridgers
Managing Director
International Affairs and Trade
Government Accountability Office
441 G Street, N.W.
Washington, D.C. 20548-0001

Dear Ms. Williams-Bridgers:

We appreciate the opportunity to review your draft report,
"PRIVACY: Opportunities Exist For Agencies and Information Resellers to
More Fully Adhere to Key Principles," GAO Job Code 310732.

The enclosed Department of State comments are provided for
incorporation with this letter as an appendix to the final report.

If you have any questions concerning this response, please contact
Brian Egan, Legal Adviser, Bureau of Legal Affairs, at (202) 647-2227.

Sincerely,

Bradford R. Higgins

cc: GAO – Jamie Pressman
      CA & DS
      State/OIG – Mark Duda

Department of State Comments on GAO Draft Report
**PRIVACY: Opportunities Exist For Agencies and Information
Resellers to More Fully Adhere to Key Principles**
(GAO-06-421 GAO Code 310732)

Thank you for giving us the opportunity to comment on GAO's draft
report "Privacy: Opportunities Exist For Agencies and Information Resellers
to More Fully Adhere to Key Principles."

In general, GAO's report seems to rest on the premise that records
from "information resellers" should be accorded special treatment when
compared with sensitive information from other sources. We do not believe
that this premise is inherently sound. The Department receives sensitive
information from a variety of sources in order to ensure that visas and
passports are issued only to those who are entitled to them, to conduct
investigations as part of its diplomatic security mission, and in other
contexts. The Department does not distinguish between types of information
or sources of information in deciding whether to comply with privacy laws.
All Department information is treated in accordance with applicable privacy
laws, regardless of the source or type of information at issue.

We also have a few specific technical comments. We request that
GAO revise those sections of the report (e.g., at 58 and 62) which suggest
that "fraud protection" in the passport and visa context is "not related to law
enforcement." The Department is charged with investigating, making
arrests, and working with other appropriate law enforcement agencies to
detect and prosecute potential cases of visa and passport fraud. In the
passport context, GAO recently stated that "[m]aintaining the integrity of the
U.S. passport is essential to the State Department's effort to protect U.S.
citizens from terrorists, criminals, and others," and that "Passport fraud is
often intended to facilitate such crimes as illegal immigration, drug
trafficking, and alien smuggling." See GAO, Improvements Needed to
Strengthen U.S. Passport Fraud Detection Efforts (June 29, 2005) at 2.
Fraud detection in the passport and visa context is clearly related to law
enforcement, as well as to the vital task of providing homeland security.

On a related note, we disagree with GAO's criticism (at 62-63) of the
use of terms such as "public source material" to identify categories of
sources of records in Privacy Act systems of records notices. To the extent
that an agency's system of record notices properly identify "categories" of
records, the notices are in compliance with the Privacy Act. See 5 U.S.C. §
552a(e)(4)(I). In our view, it would be bad policy to require separate and
specific mention of information from individual sources such as data
resellers, as this would imply that such information could not be considered
when it was **not** specifically mentioned. Such a policy could result in
critical information not being considered in a given case (in the case of the
Department, for example, in adjudicating a visa or passport application),
with consequent harmful effects on the United States national interest. The
proliferation of such requirements for "specific mention" in systems of
records notices would likely compound this problem, with the result that
USG judgments would be less, not more, well-founded.

In: Future of the Internet: Social Networks...
Editors: Rick D. Sullivan and Dominick P. Bartell
ISBN: 978-1-61209-597-4
©2011 Nova Science Publishers, Inc.

Chapter 3

PERSONAL INFORMATION: KEY FEDERAL PRIVACY LAWS DO NOT REQUIRE INFORMATION RESELLERS TO SAFEGUARD ALL SENSITIVE DATA[*]

United States Government Accountability Office

WHAT GAO FOUND

Financial institutions such as banks, credit card companies, securities firms, and insurance companies use personal data obtained from information resellers to help make eligibility determinations, comply with legal requirements, prevent fraud, and market their products. For example, lenders rely on credit reports sold by the three nationwide credit bureaus to help decide whether to offer credit and on what terms. Some companies also use reseller products to comply with PATRIOT Act rules, to investigate fraud, and to identify customers with specific characteristics for marketing purposes.

GAO found that the applicability of the primary federal privacy and data security laws— the Fair Credit Reporting Act (FCRA) and Gramm-LeachBliley Act (GLBA)—to information resellers is limited. FCRA applies to information collected or used to help determine eligibility for such things as credit or insurance, while GLBA only applies to information obtained by or from a GLBA-defined financial institution. Although these laws include data security provisions, consumers could benefit from the expansion of such requirements to all sensitive personal information held by resellers.

The Federal Trade Commission (FTC) is the primary federal agency responsible for enforcing information resellers' compliance with FCRA's and GLBA's privacy and security provisions. Since 1972, the agency has initiated formal enforcement actions against more than 20 resellers, including the three nationwide credit bureaus, for violating FCRA. However, FTC does not have civil penalty authority under the privacy and safeguarding provisions of

---

[*] This is an edited, reformatted and augmented edition of a United States Government Accountability Office publication, Report GAO-06-674, dated June 2006.

GLBA, which may reduce its ability to enforce that law most effectively against certain violations, such as breaches of mass consumer data.

In overseeing compliance with privacy and data security laws, federal banking and securities regulators have issued guidance, conducted examinations, and taken formal and informal enforcement actions. A recent national survey sponsored by the National Association of Insurance Commissioners (NAIC) identified some noncompliance with GLBA by insurance companies, but state regulators have not laid out clear plans with NAIC for following up to ensure these issues are adequately addressed.

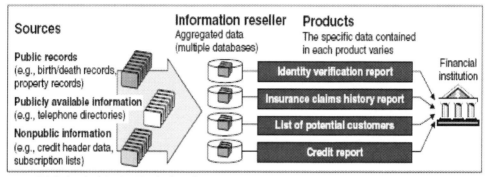

Sources: GAO (analysis); Art Explosion (images).

Typical Information Flow through Resellers to Financial Institutions.

WHY GAO DID THIS STUDY

The growth of information resellers—companies that collect and resell publicly available and private information on individuals—has raised privacy and security concerns about this industry. These companies collectively maintain large amounts of detailed personal information on nearly all American consumers, and some have experienced security breaches in recent years.

GAO was asked to examine (1) financial institutions' use of resellers; (2) federal privacy and security laws applicable to resellers; (3) federal regulators' oversight of resellers; and (4) regulators' oversight of financial institution compliance with privacy and data security laws. To address these objectives, GAO analyzed documents and interviewed representatives from 10 information resellers, 14 financial institutions, 11 regulators, industry and consumer groups, and others.

WHAT GAO RECOMMENDS

Congress should consider (1) requiring information resellers to safeguard all sensitive personal information they hold, and (2) giving FTC civil penalty authority for enforcement of GLBA's privacy and safeguarding provisions. GAO also recommends that state insurance regulators ensure compliance with GLBA.

ABBREVIATIONS

CRA	consumer reporting agency
DISB	District of Columbia's Department of Insurance, Securities and Banking
FACT Act	Fair and Accurate Credit Transactions Act
FCRA	Fair Credit Reporting Act
FDIC	Federal Deposit Insurance Corporation
FFIEC	Federal Financial Institutions Examination Council
FRB	Board of Governors of the Federal Reserve System
FTC	Federal Trade Commission
FTC Act	Federal Trade Commission Act
GLBA	Gramm-Leach-Bliley Act
NAIC	National Association of Insurance Commissioners
NCUA	National Credit Union Administration
NYSE Regulation	New York Stock Exchange Regulation
OCC	Office of the Comptroller of the Currency
OFAC	Office of Foreign Assets Control
OTS	Office of Thrift Supervision
SEC	Securities and Exchange Commission
USA PATRIOT ACT	Uniting and Strengthening America by Providing Appropriate Tools Required to Intercept and Obstruct Terrorism Act

June 26, 2006

The Honorable Richard C. Shelby
Chairman
The Honorable Paul S. Sarbanes
Ranking Minority Member
Committee on Banking, Housing and Urban Affairs United States Senate

The growth in recent years of information resellers—companies that collect, aggregate, and resell publicly available and private information on individuals—has raised privacy and security concerns related to this industry. [1] Information resellers maintain and sell vast amounts of detailed personal information on nearly all American consumers—including such things as Social Security numbers, home and automobile values, occupations and hobbies. In addition, security breaches at some of these companies have raised concerns in light of the increasing problem of identity theft. Some policymakers and consumer advocates believe that not enough is known about these resellers and the information about consumers that they maintain and share.

Information resellers include consumer reporting agencies (CRA), which assemble and share credit histories and other personal information used to help make important decisions about individuals, such as their eligibility for financial services. Other companies, sometimes called "data brokers," collect personal information from a variety of sources for such things as marketing and fraud prevention. Advances in technology and the computerization of public records in recent years have fostered significant growth in the size of the reseller industry and the amount of personal consumer data that these companies assemble and distribute.

The primary federal laws governing the sharing and use of personal information by private sector companies are the Fair Credit Reporting Act (FCRA) and subtitle A of title V of the Gramm-Leach-Bliley Act (GLBA). [2] Several federal and state agencies and self-regulatory organizations enforce these laws, including the Federal Trade Commission (FTC); the banking regulators—Board of Governors of the Federal Reserve System (FRB), Office of the Comptroller of the Currency (OCC), Office of Thrift Supervision (OTS), Federal Deposit Insurance Corporation (FDIC), and National Credit Union Administration (NCUA); the securities regulators— Securities and Exchange Commission (SEC), NASD (formerly known as the National Association of Securities Dealers), and New York Stock Exchange Regulation (NYSE Regulation); and state insurance regulators.

Concerned about financial institutions' use of information resellers, you asked us to examine (1) how financial institutions use data products supplied by information resellers, the types of information contained in these products, and the sources of the information; (2) how federal laws governing the privacy and security of personal data apply to information resellers, and what rights and opportunities exist for individuals to view and correct data held by resellers; (3) how federal financial institution regulators and the FTC oversee information resellers' compliance with federal privacy and information security laws; and (4) how federal financial institution regulators, state insurance regulators, and the FTC oversee financial institutions' compliance with federal privacy and information security laws governing consumer information, including information supplied by information resellers.

To address these objectives, we gathered and analyzed documents, and interviewed representatives from, 10 major information resellers; 14 financial institutions in the banking, securities, credit card, property/casualty insurance, and consumer lending industry sectors; and trade associations representing these firms. We also met with experts in the area of privacy law and with consumer advocacy organizations active in the field. Our audit work allows us to represent how financial institutions that offer a sizable and diverse portion of financial services in the United States use information resellers, and to describe the types of information products offered by the information resellers most commonly identified by these financial institutions. Our findings, however, are not representative of all financial institutions and information resellers. We also analyzed relevant laws, guidance, and regulations. Finally, to describe federal and state enforcement and supervisory activities, we interviewed and analyzed documents from FTC; the five federal banking and three securities regulators; the National Association of Insurance

Commissioners (NAIC), which represents state insurance regulators; and the District of Columbia's Department of Insurance, Securities and Banking (DISB).

We conducted our review from June 2005 through May 2006 in accordance with generally accepted government auditing standards. A more extensive discussion of our scope and methodology appears in appendix I.

RESULTS IN BRIEF

Financial institutions use data from information resellers to help determine individuals' eligibility for credit and insurance, comply with legal requirements, prevent fraud, and market products. Banks and other lenders use reseller data to help make eligibility and interest rate decisions for new applicants and existing customers, while insurance companies use these data to help make underwriting decisions regarding individual insurance applications. To meet PATRIOT Act requirements designed to prevent money laundering and transactions with known criminals, some financial institutions we spoke with use resellers to confirm the identity of applicants. In addition, reseller data are used to identify and investigate fraud, locate holders of delinquent accounts, and conduct due diligence on individuals associated with new business ventures. Many companies also use certain information reseller products for marketing purposes—such as to target potential customers who have certain characteristics or to gather additional information about existing customers to offer additional products. The specific information maintained by resellers varies depending on the nature of the reseller and the types and purposes of its products. Their products often include credit header data—identifying information at the top of a credit report that includes such things as name, current and prior addresses, telephone number, and Social Security number. Products used by lenders for eligibility determinations typically also contain detailed credit histories and scores, while products used by insurers may also contain past insurance claims filed by applicants. Many reseller products, particularly those used for fraud detection, include court and property records and bankruptcy filings, motor vehicle records, names of family members and associates, and professional licenses. Products used for marketing often include demographic information as well as information on individual consumers' interests and hobbies. Resellers' sources vary depending on the product, but may include public records from government agencies, publicly available information, such as telephone or business directories, and nonpublic or proprietary information from credit bureaus or provided to businesses directly by consumers.

The primary federal privacy and data security laws that apply to information resellers are the Fair Credit Reporting Act (FCRA) and the Gramm-Leach-Bliley Act (GLBA), but the applicability of these laws with regard to information resellers is limited. FCRA requires companies to safeguard and restrict their use and distribution of consumer information collected or used to determine eligibility for such things as credit, insurance, or employment, and provides rights to consumers to view and rectify errors in databases containing such information. The applicability of FCRA depends largely on the purpose for which the information is collected, and its intended and actual use, rather than the origins or nature of the information itself. Resellers offer many products from databases they consider not subject to FCRA, such as those used for many marketing and anti-fraud products. Information resellers vary in the extent to which they voluntarily provide consumers additional opportunities to view, correct, and opt out of the sharing of information that is not subject to FCRA. GLBA's privacy provisions restrict the sharing of nonpublic personal information collected by or acquired from financial institutions, except in certain circumstances. However, these provisions only apply to information resellers covered by GLBA's definition of a "financial institution" or that maintain nonpublic personal information originating from such a financial institution. GLBA's safeguarding provisions require that steps be taken to ensure the security and confidentiality of customers' nonpublic personal information, but similarly this applies only to resellers that are GLBA financial institutions. Because of the limited applicability of FCRA and GLBA to information resellers, sensitive personal information these companies maintain is often not covered by explicit statutory safeguarding requirements. For example, some information resellers maintain data such as Social Security numbers in anti-fraud databases or household incomes in marketing databases that they do not consider subject to FCRA's or GLBA's safeguarding provisions. Requiring information resellers to take steps to prevent unauthorized access to all of the sensitive personal information they hold would help ensure that explicit data security requirements apply more comprehensively to a class of companies that maintains large amounts of such data. In addition, no federal statute requires companies to disclose breaches of sensitive personal information, although such a requirement could provide incentives to companies to improve data safeguarding and provide consumers at risk of identity theft or other related harm with useful information.

FTC is the primary federal agency responsible for enforcing information resellers' compliance with the privacy and information security requirements of FCRA and GLBA. Because it is a law enforcement agency, as opposed to a regulatory or supervisory agency, FTC does not routinely monitor or examine resellers, but can initiate investigations based on complaints and other sources. Since 1972, the agency has initiated formal enforcement actions against more than 20 consumer reporting agencies, including the three nationwide credit bureaus, for violating FCRA and the Federal Trade Commission Act (FTC Act). For example, in January 2006, ChoicePoint agreed to pay $10 million in civil penalties and $5 million for consumer redress (damages to compensate consumers for losses) to settle FTC charges that the company's security and record-handling procedures allegedly violated FCRA and the FTC Act. Many of FTC's cases involved companies alleged to have provided consumer report information without adequately ensuring that their customers had a permissible purpose for obtaining it. FTC cannot impose civil penalties for violations of GLBA's privacy and safeguarding provisions, as it can under FCRA. FTC has used its existing enforcement authority under GLBA to seek injunctions against financial institutions that have violated that law, and it can also seek redress for consumers. However, FTC staff have said that civil penalties would be a more effective tool for violations involving breaches of mass consumer data.

Federal and state regulators vary in the actions they take to oversee financial institutions' compliance with federal privacy and information security laws. In general, regulators told us that their oversight activities focus on the protection of all sensitive data; they do not typically distinguish whether the data were obtained from an information reseller or some other source. The five federal banking regulators have implemented and enforced GLBA and FCRA by issuing regulations and guidance, by using their examination procedures to check compliance with these laws, and by taking enforcement actions to address violations. SEC has issued regulations to implement GLBA for broker-dealers, investment companies, and SEC-registered investment advisers. SEC, NASD, and NYSE Regulation have also issued guidance and examined securities firms for compliance with GLBA's privacy and safeguarding provisions, and as necessary have taken enforcement actions. State insurance regulators are responsible for enforcing GLBA for their states' property-casualty insurers. NAIC told us that state insurance regulators do not typically focus in their examinations on privacy requirements, but that they did recently participate in a multistate survey of insurance company compliance with GLBA. The survey identified a number of areas of noncompliance with GLBA, but the extent to which state regulators will be addressing these problems is unclear. FTC enforces securities firms' and insurance companies' compliance with FCRA and enforces both FCRA and GLBA for all financial institutions not otherwise supervised by another regulator. FTC has issued regulations to implement GLBA and initiated enforcement actions against consumer finance companies for not ensuring the security and confidentiality of sensitive customer information. Some federal banking regulators have authority to examine third-party service providers with which the banks may do business, and regulators have examined a limited number of information resellers under this authority.

This chapter suggests that Congress consider requiring information resellers, and potentially a broader class of entities, to safeguard all sensitive personal information they hold. We also suggest that Congress consider providing FTC with civil penalty authority for its enforcement of GLBA's privacy and safeguarding provisions. In addition, we recommend that state insurance regulators, individually and in concert with NAIC, take additional measures to ensure appropriate enforcement of insurance companies' compliance with GLBA's privacy and safeguarding requirements. We provided a draft of this chapter to FDIC, FRB, FTC, NAIC, NASD, NCUA, NYSE Regulation, OCC, OTS, and SEC, which provided technical comments that were incorporated as appropriate. In addition, FTC provided written comments, in which the agency noted that it agreed with our suggestions to Congress.

BACKGROUND

"Information reseller" is an umbrella term used to describe a wide variety of businesses that collect and aggregate personal information from multiple sources and make it available to their customers. The industry has grown considerably over the past two decades, in large part due to advances in computer technology and electronic storage. Courthouses and other government offices previously stored personal information in paper- based public records that were relatively difficult to obtain, usually requiring a personal visit to inspect the records. Nonpublic information, such as personal information contained in product registrations or insurance applications was also generally inaccessible. In recent years, however, the electronic storage of public and private records along with increased computer processing speeds and decreased data storage costs have fostered information reseller businesses that collect, organize, and sell vast amounts of personal information on virtually all American consumers.

The information reseller industry is large and complex, and these businesses vary in many ways. What constitutes an information reseller is not always clearly defined and little data exist on the total number of firms that offer information products. FTC and other federal agencies do not keep comprehensive lists of companies that resell personal information, and experts say that characterizing the precise size and nature of the information reseller industry can be difficult because it is evolving and lacks a clear definition. Although no comprehensive data exist, industry representatives say there are at least hundreds of information resellers in total, including some companies that provide services over the Internet. [3]

We include in our definition of information resellers the three nationwide credit bureaus—Equifax, Experian, and TransUnion, which primarily collect and sell information about the creditworthiness of individuals—as well as other resellers such as ChoicePoint, Acxiom, and LexisNexis, which sell information for a variety of purposes, including marketing. [4] Other companies that sell information products include eFunds, which provides depository institutions with information on deposit account histories; Thompson West and Regulatory DataCorp, which help companies mitigate fraud and other risks; and ISO, which provides insurers with insurance claims histories and fraud prevention products. Information resellers sell their products to a broad spectrum of customers, including private companies, individuals, law enforcement bureaus and other government agencies. [5] Although major information resellers generally offer their products only to customers who have successfully completed a credentialing process, some resellers offer certain products, such as compilations of telephone directory information, to the public at large. All of these businesses differ in nature, and they do not all focus exclusively on aggregating and reselling personal information. For example, Acxiom primarily provides customized computer services, and its information products represent a relatively small portion of the overall activities of the company.

Information resellers obtain their information from many different sources (see figure 1). Generally, three types of information are collected: public records, publicly available information, and nonpublic information.

- Public records are a primary source of information about consumers, available to anyone, and can be obtained from governmental entities. What constitutes public records is dependent upon state and federal laws, but generally these include birth and death records, property records, tax lien records, voter registrations, licensing records, and court records (including criminal records, bankruptcy filings, civil case files, and legal judgments).
- Publicly available information is information not found in public records but nevertheless publicly available through other sources. These sources include telephone directories, business directories, print publications such as classified ads or magazines, Internet sites, and other sources accessible by the general public.
- Nonpublic information is derived from proprietary or nonpublic sources, such as credit header data, product warranty registrations, lists of magazine or catalog subscribers, and other application information provided to private businesses directly by consumers. [6]

Information resellers hold or have access to databases containing a large variety of information about individuals. Although each reseller varies in the specific personal information it maintains, it can include names, aliases, Social Security numbers, addresses, telephone numbers, motor vehicle records, family members, neighbors, insurance claims, deposit account histories, criminal records, employment histories, credit histories, bankruptcy records, professional licenses, household incomes, home values, automobile values, occupations, ethnicities, and hobbies.

The various products offered by different types of information resellers are used for a wide range of purposes, including credit and background checks, fraud prevention, and marketing. Resellers often sell their data to each other—for example, the credit bureaus sell credit header data to other resellers for use in identity verification and fraud prevention products. Resellers might also purchase publicly available information from one another, rather than gathering the information themselves. The nature of the databases maintained and products offered by information resellers vary. Credit bureaus maintain an individual file on most Americans containing financial information related to that person's creditworthiness. Most other resellers do not typically maintain complete files on individuals, but rather collect and maintain information in a variety of databases, and then provide their customers with a single consolidated source for a broad array of personal information.

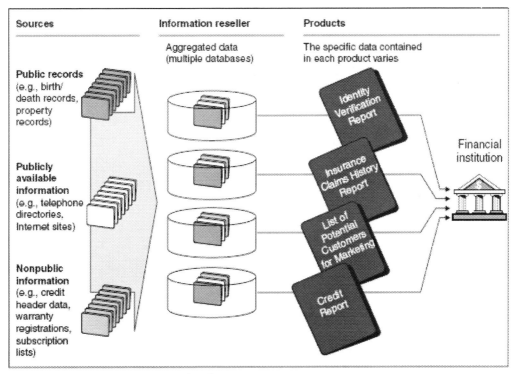

Sources: GAO (analysis); Art Explosion (images).

Figure 1. Typical Information Flow through Resellers to Financial Institutions.

FINANCIAL INSTITUTIONS USE INFORMATION RESELLERS FOR ELIGIBILITY DETERMINATIONS, FRAUD PREVENTION, PATRIOT ACT COMPLIANCE, AND MARKETING

Financial institutions in the banking, credit card, securities, and insurance industries use personal data purchased from information resellers primarily to help make eligibility determinations, comply with legal requirements, prevent fraud, and market their products. [7] Credit reports from the three nationwide credit bureaus help lenders determine eligibility for and the cost of credit, and reports on insurance claims histories from specialty CRAs help insurance companies make premium decisions for new applicants and existing customers.

To meet certain legal requirements and detect and prevent fraud, financial institutions we studied also use reseller products to locate individuals or confirm their identity. In addition, certain reseller products containing demographic data and information on individuals' lifestyle interests and hobbies are used to help market financial products to existing or potential customers with certain characteristics.

Consumer Reports Sold by Credit Bureaus and Other CRAs Are Used to Make Credit and Insurance Eligibility Decisions

Banks, credit card companies, and other lenders rely on credit reports sold by the three nationwide credit bureaus—Equifax, Experian, and TransUnion—when deciding whether to offer credit to an individual, at what rate, and on what terms. Banks use credit reports to help assess the credit risk of new customers before opening a new deposit account or providing a mortgage or other loan. Credit card companies use credit reports to determine whether to grant a credit card to an applicant, determine the terms of that card, and to adjust the account terms of current cardholders whose creditworthiness may have changed. In addition to lenders, insurance companies often use scores generated from credit report information to help determine premiums for the policies they underwrite.

Credit bureaus receive the information in credit reports from the financial institutions themselves, among other sources. Credit reports consist of a "credit header"— identifying information such as name, current and previous addresses, Social Security number, and telephone number—and a credit history, or other payment history, designed to provide information on the individual's creditworthiness. The credit history might contain information on an individual's current and past credit accounts, including amounts borrowed and owed, credit limits, relevant dates, and payment histories, including any record of late payments. Credit reports also may include public record information on tax liens, bankruptcies, and other court judgments related to the payment of debts. Credit bureaus also sell credit scores, which are numerical representations of predicted creditworthiness based on information in credit reports, and are often used instead of full credit reports. For example, all three credit bureaus sell FICO® credit scores, which use factors such as payment history, amount owed, and length of credit history to help financial institutions predict the likelihood that a person will repay a loan. [8]

Some financial institutions also use specialty CRAs, which maintain specific types of files on consumers, to help make eligibility decisions. Insurance companies commonly use products from ChoicePoint and ISO, which compile data from insurance companies on the claims that individuals have made against their homeowner's or automobile insurance policies. [9] Most insurance companies provide these CRAs with claim and loss information about their customers, including names, driver's license information, type of loss, date of loss, and amount the insurance company paid to settle the claim. The CRAs aggregate this information from multiple insurance companies to create either full reports or risk scores designed to help assess the likelihood that an individual will file a claim. Insurance companies purchase reports, or in some cases scores, associated with individuals applying for insurance and the property being insured to help decide whether to provide coverage and at what rate. Insurance companies also use this information to help determine whether to extend coverage and set premiums for existing policy holders. (See app. II for a sample insurance claims history report.) Insurance industry representatives told us aggregated claims data provided by specialty CRAs are extremely useful in making coverage and rate determinations. They noted, for example, that past losses are the best indicator of future driving risk and thus are useful to firms that underwrite auto insurance.

Banks and credit unions frequently assess applicants of new checking and other deposit accounts using products offered by resellers such as ChexSystems, a specialty CRA that is a subsidiary of eFunds. ChexSystems compiles information from banks and credit unions on accounts that have been closed due to account misconduct such as overdrafts, insufficient funds activity, returned checks, bank fraud, and check forgery. The company also aggregates available driver's license information from state departments of motor vehicles, and receives information from check- printing companies on check order histories, which can help identify fraud. Banks we spoke with said that the name and identifying information of a customer seeking to open a new deposit account is typically run through the ChexSystems database. The reports provided back to the financial institution by ChexSystems typically include identifying information, as well as information useful in assessing an applicant's risk, such as the applicant's history of check orders and the source and details of any account misconduct. (See app. II for a sample deposit account history report.)

Financial Institutions Use Information Resellers to Comply with the PATRIOT Act, Prevent Fraud, Mitigate Risk, and Locate Individuals

Financial institutions use data purchased from information resellers to comply with legal requirements; detect, prevent, and investigate fraud; identify risks associated with prospective clients; and locate debtors or shareholders.

Complying with PATRIOT Act Requirements

Financial institutions we spoke with frequently use products provided by information resellers to comply with PATRIOT Act requirements. [10] Congress intended these provisions to help prevent terrorists and other criminals from using the U.S. financial system to fund terrorism and launder money. The act requires financial institutions to develop procedures to assure the identity of new customers. [11] Many resellers offer products that verify and validate a new customer's identity by comparing information the customer provided to the financial institution with information aggregated from public and private sources. Some financial institutions, particularly those that offer services by telephone, mail, or the Internet, often confirm customers' identities using these reseller products. Other companies may verify their customers' identity from a driver's license, passport, or other paper document, but use information resellers for additional verification.

Financial institutions must also screen their customers to ensure they are not on the Department of the Treasury's Office of Foreign Assets Control (OFAC) Specially Designated Nationals and Blocked Persons List. The list includes individuals and entities that financial institutions are generally prohibited from conducting transactions with because they have been identified as potential terrorists, money launderers, international narcotics traffickers, or other criminals. Many information resellers offer products to financial institutions that screen new customers against the OFAC list; often this screening is packaged with identity verification in a single product. (See app. II for a sample identity verification and OFAC screening report.) The OFAC list is a publicly available government document, but financial institutions told us they use resellers for their screening because it allows them to do so more quickly and helps distinguish between common names on the list that might result in false matches. Some financial institutions use resellers to screen new customers against the OFAC list, while others periodically screen all of their existing customers. Some companies told us they do most of their OFAC screening internally, but sometimes use a reseller to gather additional information confirming whether a potential match is indeed an individual that is on the OFAC list.

To verify a customer's identity or conduct an OFAC screening, a financial institution typically uses a Web-based portal to provide an information reseller with basic information about the individual being screened—such as the person's name, Social Security number, address, driver's license number, phone number, and date of birth. The reseller then checks the information against its own records, and typically provides a "pass" response if the information matches, or a "fail" response if, for example, the date of birth does not match the name. Resellers' screening products generally draw on credit header data purchased from the credit bureaus, along with publicly available data such as address and telephone records and drivers' license records from state agencies. Customer verification databases also include information that may indicate suspicious activity, such as prison or campground addresses, disconnected telephone numbers, and Social Security numbers of deceased individuals.

Preventing and Detecting Fraud

The financial institutions we reviewed use information reseller tools to assist their fraud prevention and detection efforts. For example, banks and credit card companies sometimes use information reseller products to authenticate the identity of existing customers who call to update or receive account information or to order a replacement credit card. Authentication products usually draw on information similar to that used for verification products, most commonly credit header data and public records. Some resellers offer products that also allow the financial institution to access the customers' credit history with their permission, which provides additional personal information that can be used to verify identity. For example, a customer might be asked the year an automobile loan was originated or the credit limit on a credit card.

Fraud departments of financial institutions in our review also use more detailed products from information resellers to investigate suspected identity theft or account fraud, such as the use of a stolen credit card number. (See app. II for a sample fraud investigation report.) In these cases, a company's fraud department often purchases from information resellers detailed background information on a suspect's current and prior residences, vehicles, relatives, aliases, criminal records (in certain states), and other information that can be useful in directing an investigation. Examples of the uses of fraud products offered by resellers include

- obtaining detailed personal information about people associated with potential fraud, or their relatives and associates;

- detecting links between individuals who may be co-conspirators in fraud or misconduct;
- identifying multiple insurance claims made by the same person;
- identifying individuals who are associated with multiple addresses, telephone numbers, or vehicles in ways that indicate potential fraud;
- obtaining contact information for key individuals, such as witnesses to car accidents identified in police reports; or
- identifying instances where insurance policy applicants have failed to disclose certain required information.
-

Reducing Risk and Locating Individuals

Financial institutions also sometimes use reseller products to help identify potential reputational risk or other risks associated with new customers or business partners. For example, securities firms told us they screen individuals like prospective wealth management clients or merger partners to check for a criminal record, disciplinary action by securities regulators, negative news media coverage, and known affiliation with terrorism, drug trafficking, or organized crime.

Financial institutions we spoke with also often use information resellers to locate individuals. For example, lenders use reseller products to find customers who have defaulted on debts, and some mutual fund companies use these products to locate lost shareholders. The information provided by products used for this purpose is derived largely from credit header data, telephone records, and public records data, and may include an individual's aliases, addresses, telephone numbers, Social Security number, motor vehicle records, as well as the names of neighbors and associates. For example, one financial institution told us its debt collectors use a ChoicePoint product called DEBTOR Discovery to get such information to help locate delinquent debtors.

Some Financial Institutions Use Information Resellers for Marketing

Some information resellers offer certain products that help financial institutions market their financial products and services to new or existing customers with specific characteristics. Databases held by resellers offering marketing products include a variety of information on individuals and households, such as household size, number and ages of children, estimated household income, homeownership status, demographic data, and lifestyle interests and activities. These databases derive their information from public records as well as nonpublic sources such as self-reported marketing surveys, product warranty cards, and lists of magazine subscribers, which may be used to provide financial institutions and other companies with lists of consumers meeting certain criteria. [12] For example, a bank marketing a college savings account might request the names and addresses of all households in certain ZIP codes that have children under the age of 18 and household incomes of $100,000 or more. Financial institutions we studied also use certain reseller products to gather additional information on their existing customers to market additional products and services. For example, we spoke with an insurance company that used an information reseller to learn which of its existing customers owned boats, so those customers could be targeted for boat insurance. Similarly, one bank we spoke with used an information reseller to help market a sailing credit card to current customers who lived near bodies of water.

Many companies that solicit new credit card accounts and insurance policies use nationwide credit bureaus for "prescreening" to identify potential customers for the products they offer. [13] A lender or insurance company establishes criteria, such as a minimum credit score, and then purchases from a credit bureau a list of people in the bureau's database who meet those criteria. In some cases, the financial institution already has a list of potential customers that it provides to the credit bureau to identify individuals on the list who meet the criteria. Financial institutions sometimes also use a second information reseller to help them obtain from a credit bureau a list that includes only consumers meeting specific demographic or lifestyle criteria. For example, in marketing a home equity line of credit, a lender may use a second information reseller to work with a credit bureau to identify creditworthy individuals that are also homeowners and live in certain geographic areas, to which the lender will then make a firm offer of credit. Financial institutions sometimes use data from information resellers for models—developed by either the institution or the reseller—that seek to predict consumers likely to be interested in a new product and unlikely to present a credit risk. For example, a firm we spoke with that was marketing credit cards to college students used reseller data to determine the characteristics of college students that indicate they will be successful credit card borrowers.

FEDERAL PRIVACY AND INFORMATION SECURITY LAWS APPLY TO MANY INFORMATION RESELLER PRODUCTS, DEPENDING ON THEIR USE AND SOURCE

The Fair Credit Reporting Act (FCRA) and the Gramm-Leach-Bliley Act (GLBA) are the primary federal laws governing the privacy and security of personal data collected and shared by information resellers. FCRA limits resellers' use and distribution of personal data, and allows consumers to access the data held on them, but it only applies to information collected or used primarily to make eligibility determinations. Unless FCRA applies to a product and its database, resellers typically provide only limited opportunities for the consumer to access, correct, or restrict sharing of the personal data held on them. GLBA's privacy provisions restrict the sharing of nonpublic personal information collected by or acquired from financial institutions, including resellers covered by GLBA's definition of financial institution (GLBA financial institutions). Further, GLBA's safeguarding provision requires resellers that are GLBA financial institutions to safeguard this information.

Several Federal Privacy and Security Laws Apply to Personal Data Held by Information Resellers

No single federal law governs the use or disclosure of all personal information by private sector companies. Similarly, there are no federal laws designed specifically to address all of the products sold and data maintained by information resellers. [14] Instead, a variety of different laws govern the use, sharing, and protection of personal information that is maintained for specific purposes or by specific types of entities. The two primary federal laws that protect personal information maintained by private sector companies are FCRA and GLBA. FCRA protects the security and confidentiality of personal information that is collected or used to help make decisions about individuals' eligibility for, among other things, credit, insurance, or employment, while GLBA is designed to protect personal financial information that individuals provide to or that is maintained by financial institutions.

In addition to FCRA and GLBA, other federal laws that directly or indirectly address privacy and data security may also cover some information reseller products. [15] The Driver's Privacy Protection Act of 1994 regulates the use and disclosure by state motor vehicle departments of personal information from motor vehicle records. [16] Personal motor vehicle records may be purchased and sold only for certain purposes—such as insurance claims investigations and other anti-fraud activities—unless a state motor vehicle agency has received express consent from the individual indicating otherwise. [17] In addition, the Federal Trade Commission Act (FTC Act), enacted in 1914 and amended on numerous occasions, gives FTC the authority to prohibit and act against unfair or deceptive acts or practices. [18] The failure by a commercial entity, such as an information reseller, to reasonably protect personal information could be a violation of the FTC Act if the company's actions constitute an unfair or deceptive act or practice. Finally, some federal banking regulators have authority to oversee their institutions' third-party service providers to ensure the safety and soundness of financial institutions. [19] For example, if a vendor such as an information reseller did not employ reasonable safeguards to maintain a bank's records, federal banking regulators could examine the vendor to identify and remedy the risks. [20]

FCRA Applies Only to Consumer Information Used to Determine Eligibility

The Fair Credit Reporting Act (FCRA), enacted in 1970, protects the confidentiality and accuracy of personal information used to make certain types of decisions about consumers. Specifically, FCRA applies to companies that furnish, contribute to, or use "consumer reports"—reports containing information about an individual's personal and credit characteristics used to help determine eligibility for such things as credit, insurance, employment, licenses, and certain other benefits. [21] Businesses that evaluate consumer information or assemble such reports for third parties are known as consumer reporting agencies, or CRAs. Consumer reports covered by FCRA comprise a significant portion of consumer data transactions in the United States. For example, according to an industry association that represents CRAs, the three nationwide credit bureaus sell over 2.5 billion credit reports each year on average. FCRA places certain restrictions and obligations on CRAs that issue these reports. For example, the law restricts the use of consumer reports to certain permissible purposes, such as approving credit, imposes certain disclosure requirements, and requires that CRAs take steps to ensure that information in these reports is not misused. It also provides consumers with certain rights in relation to their credit reports, such as the right to dispute the accuracy or completeness of items in the reports. Congress has amended FCRA a number of times, most recently with the Fair and Accurate Credit Transactions Act of 2003 (FACT Act), which sought to promote more-accurate credit reports and expand consumers' access to their credit information. [22]

Information resellers are subject to FCRA's requirements only with regard to information used to compile consumer reports—that is, reports used to help determine eligibility for certain purposes, including credit, insurance, or employment. Thus, FCRA applies to databases used to compile credit reports sold by the three nationwide credit bureaus, and its provisions apply both to the credit bureaus themselves as well as to other information resellers that purchase and resell credit reports for use by others. FCRA also applies to databases used to generate specialty consumer reports— which consist of such things as tenant history, check writing history, employment history, medical information, or insurance claims—that are used to help make eligibility determinations. For example, according to ChoicePoint, FCRA applies to the data used in most of its WorkPlace Solutions products, which employers use to make hiring decisions. Similarly, according to LexisNexis, FCRA applies to its Electronic Bankruptcy Notifier product data, which financial institutions use to determine whether to offer customers credit or other financial services. Overall, 8 of the 10 information resellers we spoke with said that at least some of their products are consumer reports as defined by FCRA. They said their contracts prohibit their customers from using their non-FCRA products for purposes related to making eligibility determinations.

According to the information resellers included in our review, FCRA does not cover many databases used to create other products they offer because, as defined by the law, the information was not collected for making eligibility determinations and the products are not intended to be used for making eligibility determinations. [23] For example, some of the information resellers we spoke with did not treat data in some products used to identify and prevent fraud as subject to FCRA. Similarly, resellers do not typically consider databases used solely for marketing purposes to be covered by FCRA. Because the definition of a consumer report under FCRA depends on the purpose for which the information is collected and on the reports' intended and actual use, an information reseller apparently may have two essentially identical databases with only one of them subject to FCRA.

FCRA also restricts financial institutions and other companies that use consumer reports from using them for purposes other than those permitted in the law. Financial institutions must also notify consumers if they take an adverse action—such as denying an applicant a credit card— based on information in a consumer report. Under FCRA, companies that furnish information to CRAs also must take steps to ensure the accuracy of information they report. Further, users of consumer reports must properly dispose of consumer reports they maintain. The law also limits financial institutions and other entities from sharing certain credit information with their affiliates for marketing purposes. Final regulations to implement this statutory limitation have not yet been promulgated.

FCRA Provides Access, Correction, and Opt-Out Rights for Consumer Reports

FCRA is the primary federal law that provides rights to consumers to view, correct, or opt out of the sharing of their personal information, including data held by information resellers. Under FCRA, as recently amended by the FACT Act, consumers have the right to

- obtain all of the information about themselves contained in the files of a CRA upon request, including their credit history;
- receive one free copy of their credit file from nationwide CRAs and nationwide specialty CRAs once a year or under certain other circumstances; [24]
- dispute information that is incomplete or inaccurate, and have their claims investigated and any errors deleted or corrected, as provided by the law; and
- opt out of allowing CRAs to provide their personal information to third parties for prescreened marketing offers. [25]

Most of FCRA's access, correction, and opt-out rights apply not just to the three nationwide credit bureaus—Experian, TransUnion, and Equifax— but also to other CRAs, including nationwide specialty CRAs that provide reports on such things as insurance claims and tenant histories. The law imposes slightly different requirements on these entities with respect to free annual reports. For example, FCRA's implementing regulation requires Experian, TransUnion, and Equifax to create a centralized source for accepting consumer requests for free credit reports, which must include a single dedicated Web site, a toll-free telephone number, and mail directed to a single postal address where consumers can order credit reports from all three nationwide CRAs. [26] Nationwide specialty CRAs are individually required to maintain a toll-free number and a streamlined process for accepting and processing consumer requests for file disclosures. [27] Other CRAs must provide consumers with a copy of their report upon request (although in most cases they may charge a reasonable fee for it), and they must allow consumers to dispute information they believe to be inaccurate. In practice, consumers may find it difficult in some cases to effectively access and correct information held by nationwide specialty CRAs because there may be hundreds of such CRAs and no master list exists. For example, job seekers who want to confirm the accuracy of information about themselves in background-screening products would need to request their consumer reports from the dozens of such companies that offer such products.

Consumers generally do not have the legal right to access or correct information about them contained in non-FCRA databases, such as those used for marketing purposes or, in some cases, fraud detection. The information resellers we studied varied in the extent to which they voluntarily provide consumers with additional opportunities to view, correct, and opt out of the sharing of information beyond what the law requires. The three nationwide credit bureaus allowed consumers to view only information that is subject to FCRA. However, three other information resellers we spoke with allowed consumers to order summary reports of some data maintained about them that was not subject to FCRA. These reports varied in length and detail but typically contained consumer data obtained from public records, publicly available information, and credit header information. Consumers did not typically have the right to see data maintained about them related to marketing, such as information on their household income, interests, or hobbies, which was often obtained from warranty cards or self-reported survey questionnaires.

Information resellers told us that consumers who request correction of inaccurate data not covered by FCRA are typically referred to the government or private entity that was the source of the data. Many resellers told us that because their databases are so frequently updated, simply correcting their own databases would not be effective because it would soon be refreshed by new erroneous data from the original source. However, one reseller told us it has procedures that prevent such corrections from being overwritten. Some resellers offered limited opportunities for consumers to opt out of their databases even for data not covered by FCRA, but they typically allow this only for data used for marketing purposes. The five resellers we spoke with that maintain personal data used for marketing allowed consumers to request that their information not be shared with third parties. None of the resellers we spoke with offered all consumers the ability to opt out of identity verification or fraud products. They noted that it would undermine the effectiveness of the databases if, for example, criminals could remove themselves from lists of fraudsters. Some resellers do allow opt-out opportunities to certain individuals, such as judges or identity-theft victims, who may face potential harm from having their information included in reseller databases.

Industry representatives, consumer advocates, and others offer differing views on whether the access, correction, and opt-out rights provided under FCRA should be expanded. Many consumer advocates and others have argued that these rights should not be limited to consumer information used for eligibility purposes, but should explicitly extend as well to databases not currently considered by resellers to be subject to FCRA, such as those used for some anti-fraud products. Proponents of this view argue that basic privacy principles dictate that consumers should have the right to know what information is being collected and maintained about them. In addition, they argue that errors in these databases have the potential to harm consumers. For example, an individual could be denied a volunteer opportunity or falsely pursued as a crime suspect due to erroneous information in a reseller database not covered under FCRA.

In contrast, some information resellers, financial services firms, and law enforcement representatives have argued that providing individuals expanded access, correction, and opt-out rights is unnecessary and could harm fraud prevention and criminal investigations by providing individuals with the opportunity to see and manipulate the information that exists about them. They also note that expanding these rights could create new regulatory burdens. For example, firms maintaining databases for marketing purposes could face substantial costs and complications developing and implementing processes for consumers to see, challenge, and correct the data held on them. Information resellers noted that providing access and correction rights for personal information in marketing databases makes little sense because the accuracy of this information is much less important than for information used to make crucial eligibility decisions.

GLBA Applies to Information Resellers That Are Financial Institutions or Receive Information from Financial Institutions

The Gramm-Leach-Bliley Act (GLBA), enacted in 1999, limits with certain exceptions the sharing of consumer information by financial institutions and requires them to protect the security and confidentiality of customer information. Further, GLBA limits the reuse and redisclosure of the information for those receiving it. GLBA's key provisions with regard to information resellers, therefore, cover the privacy, reuse, redisclosure, and safeguarding of information.

GLBA Privacy Provisions

GLBA's privacy provisions generally limit financial institutions from sharing nonpublic personal information with nonaffiliated companies without first providing certain notice and, where appropriate, opt-out rights to their own customers and other consumers with whom they interact. [28] GLBA distinguishes between a financial institution's "customers" and other individuals the financial institution may interact less with, which the law refers to as "consumers." Specifically, a consumer is an individual who obtains a financial product or service from a financial institution. [29] On the other hand, a customer is a consumer who has an ongoing relationship with a financial institution. For example, someone who engages in an isolated transaction with a financial institution, such as obtaining an ATM withdrawal, is a consumer, whereas someone who has a deposit account with a bank would be a customer. While some GLBA requirements, such as the privacy requirements, apply broadly to cover consumer information in many cases, other provisions of GLBA apply only to customer information. For example, GLBA's safeguarding requirements oblige financial institutions to protect only customer information.

GLBA requires financial institutions to provide their customers with a notice at the start of the customer relationship and annually thereafter for the duration of that relationship. The notice must describe the company's sharing practices and give customers, and in some cases consumers, the right to opt out of some sharing. GLBA exempts companies from notice and opt-out requirements under certain circumstances. For example, financial institutions and CRAs may share personal information for credit- reporting purposes without providing opt-out opportunities, and financial institutions and others may also share this information to protect against or prevent actual or potential fraud and unauthorized transactions. [30] Thus, financial institutions are not required to provide their customers with opt- out rights before reporting their information to credit bureaus or sharing their information with information resellers for identity verification and fraud purposes. Under another GLBA exception, financial institutions are also not required to provide consumers with an opportunity to opt out of the sharing of information with companies that perform services for the financial institution. [31]

GLBA's privacy provisions apply to information resellers only if (1) the reseller is a GLBA "financial institution" or (2) the reseller receives nonpublic personal information from such a financial institution (see figure 2). The determination of whether a company is a financial institution under GLBA is complex and, for an information reseller, depends on whether the company's activities are included in implementing regulations issued by FTC. GLBA defines "financial institutions" as entities that are in the business of engaging in certain financial activities. [32] Such activities include, among other things, traditional banking services, activities that are financial in nature on the FRB list of permissible activities for financial holding companies in effect as of the date of GLBA's enactment, and new permissible activities. [33] While new financial activities may be identified, those activities are not automatically included in FTC's definition. [34] FTC defines "financial institutions" as businesses that are "significantly engaged" in financial activities. [35] For example, FRB's list of "financial activities" includes not only the activity of extending credit, but also related activities such as credit bureau services. [36] Thus, the three nationwide credit bureaus are considered financial institutions subject to GLBA. [37]

FTC staff told us that the determination of whether a specific information reseller is a financial institution subject to GLBA depends on the specific activities of the company. They said they determine whether GLBA applies to an entity on a case-by-case basis and that it is difficult to generalize what types of information resellers are GLBA financial institutions. For example, CRAs other than the three nationwide credit bureaus may not necessarily be subject to GLBA if, for example, their activities do not fall under FRB's definition of credit bureau services or they do not otherwise engage in any financial activity included in the 1999 FRB list. Only four resellers with whom we spoke—the three nationwide credit bureaus and a specialty CRA that collects deposit account information—told us they consider themselves financial institutions subject to GLBA's privacy and safeguarding provisions. Moreover, we were told that these provisions do not apply to the entire company but rather only to those activities of the company that are deemed financial in nature. For example, one credit bureau told us that its credit reporting activities fall under GLBA, but that its marketing products, which are not deemed financial in nature, do not fall under GLBA. [38]

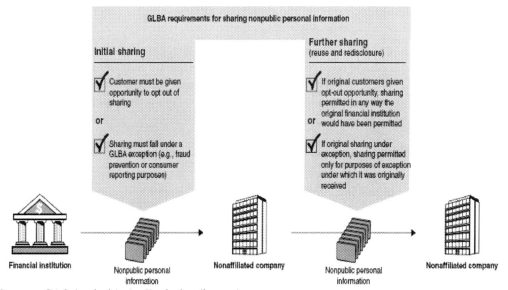

Sources: GAO (analysis); Art Explosion (images).

Figure 2. GLBA Privacy Provisions.

GLBA not only limits how financial institutions share nonpublic personal information with other companies, but it also restricts what those companies subsequently do with the information. Under GLBA's "reuse and redisclosure" provision and FTC's implementing rule, companies that receive information from a financial institution are restricted in how they further share or use that information. [39] If a company receives information under a GLBA exception, then the reseller can only reuse and redisclose the information for activities that fall under the exception under which the information was received. [40] Alternatively, if a company receives information from a financial institution in a way not covered by an exception—where an individual has been provided with a GLBA notice and has chosen not to opt out of sharing—then the information may be reused and redisclosed in any way the original financial institution would have been permitted. [41]

As noted earlier, the nationwide credit bureaus sell credit header data— identifying information at the top of a credit report—to other information resellers for use in fraud prevention products. Representatives of two of the credit bureaus and their industry association told us that because credit header data contains information from financial institutions, it is subject to GLBA's reuse and redisclosure provisions. As a result, the credit bureaus can only sell credit header data under the same GLBA exception under which they received it. Credit bureau representatives said they receive the information from financial institutions under both the consumer reporting and fraud prevention exceptions, and then sell it under the fraud prevention exception.

Also, some old credit header data may not be subject to GLBA at all. Prior to GLBA's enactment in 1999, credit header information sold by credit bureaus—which included names, addresses, aliases, and Social Security numbers—could be used or resold by a third party for any purpose, as long as the information was not used to make eligibility determinations. GLBA placed restrictions on the sale of such nonpublic personal information maintained by GLBA financial institutions. Further, as noted earlier, reuse and redisclosure of the information is also restricted by GLBA. The law's privacy restrictions generally became fully effective on July 1, 2001. [42] A nationwide credit bureau told us that the restrictions did not apply retroactively to credit header data that credit bureaus already held at the time of GLBA's enactment in 1999. The nationwide credit bureau said that just prior to GLBA's enactment, it created a new database containing "pre-GLBA" credit header data and transferred those data to a separate affiliated company. [43] The company told us that because it gathered these data prior to GLBA's enactment, the data are not subject to GLBA's privacy and safeguarding provisions.

GLBA Safeguarding Provisions

The safeguarding provisions of GLBA require financial institutions to take steps to ensure the security and confidentiality of their customers' nonpublic personal information. [44] Specifically, the agency regulations provide that financial institutions must develop comprehensive written policies and procedures to ensure the security and confidentiality of customer records and information, protect against any anticipated threats or hazards to the security or integrity of such records, and protect against unauthorized access to or use of such records or information that could result in substantial harm or inconvenience to any customer. [45] Although the privacy provisions of GLBA apply broadly to financial institutions' consumers, GLBA's safeguarding requirements only establish obligations on financial institutions to protect their customer information.

Only information resellers defined as financial institutions under the law are required to implement these safeguards. Several of the information resellers we spoke with noted that although GLBA does not apply to all of their products, they have policies and procedures to protect all of their information in a way consistent with GLBA's safeguarding requirements. Unlike GLBA's notice and opt-out requirements (privacy requirements), the law's safeguarding provisions do not directly extend to third-party companies that receive personal information from financial institutions. However, federal agencies' provisions implementing GLBA safeguarding rules require financial institutions to monitor the activities of their service providers and require them by contract to implement and maintain appropriate safeguards for customer information [46].

Many commercial entities—including many information resellers—are not subject to GLBA and therefore are not explicitly required by a federal statute to have in place policies and procedures to safeguard individuals' personal data. This raises concerns given that identity theft has emerged as a serious problem and that breaches of sensitive personal data have occurred at a variety of companies that are not financial institutions. For example, in 2005, BJ's Wholesale Club, which is not considered a GLBA financial institution, settled FTC charges that it engaged in an unfair or deceptive act or practice in violation of the FTC Act by failing to take appropriate security measures to protect the sensitive information of thousands of its customers. [47] FTC alleged that the company's failure to secure sensitive information was an unfair practice because it caused substantial injury not reasonably avoidable by consumers and not outweighed by offsetting benefits to consumers or competition. Some policymakers, consumer advocates, and industry representatives have advocated explicit statutory requirements that would expand more broadly the number and types of companies that must safeguard their data. Had there been a statutory requirement for BJ's Wholesale Club to safeguard sensitive information, FTC would have had authority to file a complaint based on the company's failure to safeguard information. Expanding the class of entities subject to safeguarding laws would impose explicit data security provisions on a larger group of organizations that are maintaining sensitive personal information. FTC has testified that should Congress enact new data security requirements, FTC's safeguards rule should serve as a model for an effective enforcement standard because it provides sufficient flexibility to apply to a wide range of companies rather than mandate specific technical requirements that may not be appropriate for all entities. [48] To be most effective, new data security provisions would need to apply both to customer and noncustomer data because the nature of information reseller businesses is such that they hold large amounts of sensitive personal information on individuals who are not their customers.

No Federal Statute Requires Notification of Data Breaches

Currently, there is no federal statute requiring information resellers or most other companies to disclose breaches of sensitive personal information, although at least 32 states have enacted some form of breach notification law. [49] Policymakers and consumer advocates have raised concerns that federal law does not always require companies to reveal instances of the theft or loss of sensitive data. These concerns have been triggered in part by increased public awareness of the problem of identity theft and by a large number of data breaches at a wide variety of public and private sector entities, including major financial services firms, information resellers, universities, and government agencies. In 2005, ChoicePoint acknowledged that the personal records it held on approximately 162,000 consumers had been compromised. As part of a settlement with the company in January 2006, FTC alleged that ChoicePoint did not have reasonable procedures to screen prospective subscribers to its data products, and provided consumers' sensitive personal information to subscribers whose applications should have raised obvious suspicions. [50] A December 2005 report by the Congressional Research Service noted that personal data security breaches were occurring with increasing regularity, and listed 97 recent breaches, five of which had occurred at information resellers. [51] Data breaches are not limited to private sector entities, as evidenced by the theft discovered in May 2006 of electronic data of the Department of Veterans Affairs containing identifying information for millions of veterans.

Congress has held several hearings related to data breaches, and a number of bills have been introduced that would require companies to notify individuals when such breaches occur. [52] The bills vary in many ways, including differences in who must be notified, the level of risk that triggers a notice, the nature of the notification, exceptions to the requirement, and the extent to which federal law preempts state law. Breach notification requirements have two primary benefits. First, they provide companies or other entities with incentives to follow good security practices so as to avoid the legal liability or public relations risks that may result from a publicized breach of customer data. Second, consumers who are informed of a breach of their personal data can take actions to mitigate potential risk, such as reviewing the accuracy of their credit reports or credit card statements. However, FTC and others have noted that any federal requirements should ensure that customers receive notices only when they are at risk of identity theft or other related harm. To require notices when consumers are not at true risk could create an undue burden on businesses that may be required to provide notices for minor and insignificant breaches. It could also overwhelm consumers with frequent notifications about breaches that have no impact on them, reducing the chance they will pay attention when a meaningful breach occurs. At the same time, consumer and privacy groups and other parties have warned against imposing too weak of a trigger for notification, and expressed concerns that a federal breach notification law could actually weaken consumers' security if it were to preempt stronger state laws. [53]

FTC HAS PRIMARY RESPONSIBILITY FOR ENFORCING INFORMATION RESELLERS' COMPLIANCE WITH PRIVACY AND INFORMATION SECURITY LAWS

The Federal Trade Commission is the federal agency with primary responsibility for enforcing applicable privacy and information security laws for information resellers. Since 1972, FTC has initiated numerous formal enforcement actions against information resellers for providing consumer report information without adequately ensuring that their customers had a permissible purpose for obtaining the data. FTC has civil penalty authority for violations of FCRA and, in limited situations, the FTC Act, but it does not have such authority for GLBA, which may inhibit its ability to most effectively enforce that law's privacy and security provisions.

FTC Has Primary Federal Enforcement Authority over Information Resellers

FTC enforces the privacy and security provisions of FCRA and GLBA over information resellers. FCRA provided FTC with enforcement authority for nearly all companies not supervised by a federal banking regulator. [54] Similarly, GLBA provided FTC with rule-making and enforcement authority over all financial institutions and other entities not under the jurisdiction of the federal banking regulators, NCUA, SEC, the Commodity Futures Trading Commission, or state insurance regulators. [55] In addition, the FTC Act provides FTC with the authority to investigate and take administrative and civil enforcement actions against most commercial entities, including information resellers, that engage in unfair or deceptive acts or practices in or affecting commerce. According to FTC officials, an information reseller could violate the FTC Act if it mishandled personal information in a way that rose to the level of an unfair or deceptive act or practice.

State regulators also play a role in enforcing data privacy and security laws. FCRA provides enforcement authority to a state's chief law enforcement officer, or any other designated officer or agency, although federal agencies have the right to intervene in any state-initiated action. [56] In addition, GLBA allows states to enforce their own information security and privacy laws, including those that provide greater protections than GLBA, as long as the state laws are not inconsistent with requirements under the federal law.

Several states, including Connecticut, North Dakota, and Vermont, have enacted restrictions on the sharing of financial information that are stricter than GLBA. [57] States can also enforce their own laws related to unfair or deceptive acts or practices to the extent the laws do not conflict with federal law.

FTC Has Investigated and Initiated Formal Enforcement Actions against Information Resellers for FCRA and FTC Act Violations

Since 1972, FTC has initiated numerous formal enforcement actions against at least 20 information resellers for violating FCRA and, in some cases, the FTC Act. [58] All of these companies were CRAs, and they included the three nationwide credit bureaus as well as a variety of types of specialty CRAs. [59] In most of these cases, FTC charged that the companies provided consumer report information without adequately ensuring that their customers had a permissible purpose for obtaining the data. In many cases, FTC alleged the companies sold consumer reports to users they had no reason to believe intended to use the information legally, or didn't require the users to identify themselves and certify in writing the purposes for which they wished to use the reports. In addition, some companies' reports allegedly included significant inaccuracies or obsolete information; some companies also failed to reinvestigate disputed information within a reasonable period of time. [60]

Among the most significant of these FTC enforcement actions against information resellers are the following:

- In 1995, FTC settled charges with Equifax Credit Information Services, the credit bureau subsidiary of Equifax Inc., for alleged violations of FCRA. FTC alleged that the company furnished consumer reports to individuals without a permissible purpose, included derogatory information in consumer reports that should have been excluded after it was disputed by the consumer, and failed to take steps to reduce inaccuracies in reports and reinvestigate disputed information. The consent agreement required Equifax to take steps to improve the accuracy of its consumer reports and limit the furnishing of such reports to those with a permissible purpose under FCRA. [61]
- In 2000, FTC ordered the TransUnion Corporation, a nationwide credit bureau, to stop selling consumer reports in the form of target marketing lists to marketers who lack an authorized purpose under FCRA for receiving them. The company had been selling mailing lists of the names and addresses of consumers meeting certain credit-related criteria (such as having certain types of loans). FTC found that the lists were

consumer reports and that the lists therefore could not be sold for target marketing purposes. [62]

- In January 2006, FTC settled charges against ChoicePoint that its security and record-handling procedures violated federal laws with respect to consumers' privacy. FTC had alleged the company violated FCRA by providing sensitive personal information to customers despite obvious indications that the information would not be used for a permissible purpose. For example, ChoicePoint allegedly approved as customers individuals who subscribed to data products for multiple businesses using fax machines in public commercial locations. FTC also charged that the company violated the FTC Act by making false and misleading statements in its privacy policy, which said it provided consumer reports only to businesses that complete a rigorous credentialing process. Under the terms of the settlement, ChoicePoint agreed to pay $10 million in civil penalties—the largest civil penalty in FTC history—and to provide $5 million in consumer redress. [63] ChoicePoint did not admit to a violation of law in settling the charges. A company representative told us it has taken steps since the breach to enhance its customer screening process and to assist affected consumers.

FTC Cannot Levy Civil Penalties for GLBA Information Privacy and Security Violations

FTC is the primary federal agency monitoring information resellers' compliance with privacy and security laws, but it is a law enforcement rather than supervisory agency. Unlike federal financial institution regulators, which oversee a relatively narrow class of entities, FTC has jurisdiction over a large and diverse group of entities and enforces a wide variety of statutes related to antitrust, financial regulation, consumer protection, and other issues. FTC's mission and resource allocations focus on conducting investigations and, unlike federal financial regulators, FTC does not routinely monitor or examine the companies over which it has jurisdiction.

If FTC has reason to believe that violations of laws under its jurisdiction have taken place, it may initiate a law enforcement action. Under its statutory authority, it can ask or compel companies to produce documents, testimony, and other materials. FTC may in administrative proceedings issue cease and desist orders for unfair or deceptive acts or practices. Further, FTC generally may seek from the United States district courts a wide range of remedies, including injunctions, damages to compensate consumers for their actual losses, and disgorgement of ill- gotten funds. [64] Depending on the law it is enforcing, FTC may also seek to obtain civil penalties—monetary fines levied for a violation of a civil statute or regulation.

Although FTC has civil penalty authority for violations of FCRA and in limited situations the FTC Act, GLBA's privacy and safeguarding provisions do not give it such authority. [65] Currently, FTC may seek an injunction to stop a company from violating these provisions and may seek redress—damages to compensate consumers for losses—or disgorgement. However, determining the appropriate amount of consumer compensation requires having information on who and how many consumers were affected and the harm, in monetary terms, that they suffered. This can be extremely difficult in the case of security and privacy violations, such as data breaches. Such breaches may lead to identity theft, but FTC staff told us that they may not be able to identify exactly which individuals were victimized and to what extent they were harmed—particularly in cases where the potential identity theft could occur years in the future. FTC could benefit from having the authority to impose civil penalties for violations of GLBA's privacy and safeguarding provisions because such penalties may be more practical enforcement tools for violations involving breaches of mass consumer data.

FTC has testified that such authority is often the most appropriate remedy in such cases, and staff told us it could more effectively deter companies from violating provisions of GLBA. Unlike FTC, other regulators have civil penalty authority to enforce violations of GLBA. For example, OCC told us it can enforce GLBA privacy and safeguard provisions with civil money penalties against any insured depository institution or institution-affiliated party. [66]

AGENCIES DIFFER IN THEIR OVERSIGHT OF THE PRIVACY AND SECURITY OF PERSONAL INFORMATION AT FINANCIAL INSTITUTIONS

In enforcing privacy and security requirements, federal regulators do not distinguish between the data that regulated entities obtain from information resellers and other personal information these entities maintain. Federal banking regulators have overseen compliance with the privacy and security provisions of GLBA and FCRA by issuing rules and guidance, conducting examinations, and taking formal and informal enforcement actions when needed. Securities and insurance regulators enforce GLBA information privacy and security requirements in a similar fashion, but FTC is responsible for FCRA enforcement among these firms. FTC is also responsible for GLBA and FCRA enforcement for financial services firms not supervised by another regulator and has initiated several enforcement actions, though it does not conduct routine examinations. Credit union, securities, and insurance regulators told us that unlike most of the banking regulators, they do not have full authority to examine their entities' third-party service providers, including information resellers.

Financial Institutions and Their Regulators Said They Do Not Distinguish between Data from Information Resellers and Other Sources

The information privacy and security provisions of GLBA and FCRA provide several federal and state agencies with authority to enforce the laws' provisions for financial institutions. As shown in figure 3, GLBA assigns federal banking and securities regulators and state insurance regulators with enforcement responsibility for the financial institutions they oversee, and FTC has jurisdiction for all other financial institutions. FCRA similarly assigns the federal banking regulators authority over the institutions they oversee and FTC with jurisdiction over other entities. [67] FCRA assigns FTC with enforcement responsibility for securities and insurance companies and provides securities and insurance regulators with no statutory responsibilities to enforce FCRA. [68]

Financial regulators told us that in their oversight of companies' compliance with privacy laws, they generally do not distinguish between data obtained from information resellers versus other sources. The nonpublic personal information maintained by financial institutions includes both data they collect directly from their customers as well as data purchased from information resellers, such as credit reports or marketing lists. Banking and securities regulators told us their efforts to oversee the privacy and security of nonpublic personal information do not focus in particular on data that came from information resellers but rather look holistically at a financial institution's information security and compliance with applicable laws. For example, OCC and FRB officials said their examiners enforce the privacy and safeguarding requirements of GLBA and FCRA regardless of whether the source of the data is an information reseller, a customer, or other source.

GLBA's safeguarding requirements apply only to nonpublic personal information that financial institutions maintain on their customers and not to information they maintain about other consumers (noncustomers). However, representatives of financial institutions we interviewed said that as a matter of policy, they generally apply the same information safeguards to both customer and consumer information. They said that their information safeguards focus on the sensitivity of the information rather than whether the person is a customer. For example, files containing Social Security numbers would have more stringent safeguards than those containing only names and addresses.

Type of financial institution	Regulator by law	
	GLBA	FCRA
National banks	OCC	
State banks (members of the Federal Reserve System)	FRB	
State banks (nonmembers of the Federal Reserve System)	FDIC	
Savings associations	OTS	
Credit unions	NCUA[a]	
Investment companies and advisers, and broker-dealers	SEC[b]	FTC[c]
Insurance companies	State regulators	FTC
Financial institutions not under other regulators' jurisdiction	FTC	

☐ Financial institutions with the same regulator under each law.

▨ Financial institutions with a different regulator under each law.

Law Source: GAO.

Notes: The Commodity Futures Trading Commission, which was not identified as a functional regulator by GLBA, is nevertheless responsible for enforcing information privacy and security requirements among futures commission merchants, commodity trading advisers, commodity pool operators, and introducing brokers subject to its jurisdiction. See 7 U.S.C. § 7b-2.

[a]NCUA enforces GLBA at all federally insured credit unions and FCRA at all federally chartered credit unions. FTC has enforcement authority for all other credit unions not subject to NCUA's jurisdiction.

[b]SEC is responsible for enforcing GLBA compliance for investment advisers registered with SEC; FTC is responsible for enforcement at all other investment advisers.

[c]FTC is responsible for enforcing FCRA at securities firms and insurance companies, but it is not a supervisory agency and does not conduct routine examinations.

Figure 3. Enforcement Responsibilities for Selected Financial Institutions under FCRA and GLBA.

Officials of a global investment banking and brokerage firm told us that although their firm maintains separate databases on customers and consumers targeted for marketing, both databases use the higher security standard required for customer information. Another company with similar practices noted that it treats all information with higher standards rather than setting up many different safeguarding policies and procedures.

Other companies noted that public relations and reputational risk concerns motivate them to maintain high safeguards to prevent any consumer information from being lost or stolen. Similarly, federal banking regulators told us that failing to safeguard consumer information may not be a violation of GLBA but is still taken very seriously because it represents a threat to a bank's safety and soundness, poses reputational risks, and reflects a weakness in a bank's corporate governance.

Federal Banking Agencies Provide Guidance and Examine Regulated Banking Organizations for GLBA and FCRA Compliance

The banking regulators responsible for GLBA and FCRA enforcement have issued regulations and other guidance on information privacy and security requirements. The individual banking regulators examine the financial institutions under their jurisdiction for compliance with GLBA and FCRA information privacy and safeguarding requirements and have taken enforcement actions for violations.

Regulations and Other Guidance

The banking agencies acting jointly and individually, and in coordination with FTC, have issued regulations and other guidance for financial institutions to follow in implementing the privacy and safeguarding requirements of GLBA. [69] In 2000, following the law's passage, the banking agencies—OCC, FRB, OTS, FDIC, and NCUA—issued rules for compliance with the law's information privacy requirements. [70] These rules helped financial institutions implement GLBA's notice and opt-out requirements. For example, they provided examples of types of information regulated by GLBA. In 2001, the agencies jointly issued guidelines establishing standards for GLBA's safeguarding requirements to assist financial institutions in establishing administrative, technical, and physical safeguards for customer information as required by law. [71] In addition to the guidelines that implement GLBA safeguarding requirements, these regulators have in some cases issued guidance to provide further assistance to their institutions. For example, the banking agencies issued a guide on small entities' compliance with GLBA's privacy provision to help companies identify and comply with the requirements. The banking agencies also have issued additional written interagency guidance for financial institutions relating to notification of their customers in the event of unauthorized access to their information where misuse of the information has occurred or is reasonably possible. [72]

The banking regulators have also issued rules and regulations for their institutions to implement certain provisions of the Fair and Accurate Credit Transactions Act of 2003 (FACT Act), which amends FCRA. [73] For example, in 2004, in coordination with FTC, these agencies issued a final rule to implement the FACT Act requirement that persons, including financial institutions, properly dispose of consumer report information and records.

[74] Some provisions—such as restrictions on how financial institutions can share data with their affiliates for marketing purposes— have yet to be finalized by the banking or other agencies.

Through the Federal Financial Institutions Examination Council (FFIEC)—a formal interagency body comprising representatives from OCC, OTS, FRB, FDIC, and NCUA that coordinates examination standards and procedures for their institutions—the banking agencies have also issued guidance to help bank examiners oversee the integrity of information technology at their institutions. For example, FFIEC developed the FFIEC IT Examination Handbook, which is composed of 12 booklets designed to help examiners and organizations determine the level of security risks at financial institutions and evaluate the adequacy of the organizations' risk management. Representatives of banking regulators say their examiners rely on these booklets in addition to the GLBA and FCRA guidance when examining the integrity of an institution's information privacy and security procedures. Some of these booklets help examiners oversee financial institutions' use of information resellers and other third-party technology service providers by addressing topics such as banks' outsourcing of technology services, or banks' supervision of its technology service providers. Financial institution regulators told us their examiners use these booklets to oversee the soundness of their institutions' technology services and to address information security issues posed by third-party technology service providers such as information resellers.

Examinations and Enforcement Actions

Banking regulators regularly examine regulated banks, thrifts, and credit unions for compliance with GLBA and FCRA requirements. [75] Each regulatory agency told us that their agencies' safety and soundness, compliance, and information technology examinations include checks on whether their institutions are in compliance with GLBA's and FCRA's provisions related to the privacy and security of personal information. For example, OCC examination procedures tell examiners to review banks' monitoring systems and procedures to detect actual and attempted attacks on or intrusions into customer information systems. However, the scope of the regulators' reviews with regard to privacy and security matters can vary depending on the degree of risk associated with the institution examined.

According to the banking agencies, their examinations of institutions' GLBA and FCRA compliance have discovered limited material deficiencies and violations requiring formal enforcement actions. Instead, they have mostly found various weaknesses that they characterized as technical in nature and required informal corrective action. [76] FDIC officials said that between 2002 and 2005, the agency took 12 formal enforcement actions for GLBA violations and no formal enforcement actions under FCRA. They noted that FDIC has also taken informal enforcement actions to correct an institution's overall compliance management system, which covers all of the consumer protection statutes and regulations in the examination scope.

According to OCC officials, between October 1, 2000, and September 30, 2005, the agency took 18 formal enforcement actions under GLBA and no formal enforcement actions under FCRA. OCC's actions in these cases resulted in outcomes such as cease and desist orders and civil money penalties levied against violators. The agency also informally required banks to take corrective action in several instances, such as requiring a bank to notify customers whose accounts may have been compromised, or requiring a bank to correct and reissue its initial privacy notice. According to OCC staff, OCC's examinations for compliance with GLBA's privacy requirements most commonly found that banks' initial privacy notices were not clear and conspicuous, and its examinations for compliance with GLBA's safeguarding requirements most commonly found cases of inadequate customer information programs, risk assessment processes, testing, and reports to the board.

FRB officials said the agency has taken 12 formal enforcement actions in the past 5 years for violations of GLBA's information-safeguarding standards and no formal actions for FCRA violations. They said FRB has taken several informal enforcement actions, including three related to violations of Regulation P, which implements GLBA's privacy requirements, and five informal actions for violations of FCRA. According to FRB staff, FRB's examinations for compliance with the interagency information security standards have found cases of inadequate customer information security programs, board oversight, and risk assessments, as well as cases of incomplete assessment of physical access controls and safeguarding of the transmission of customer data. The most commonly found problem in FRB's examinations for compliance with Regulation P was banks' failure to provide clear and conspicuous initial notices of their privacy policies and procedures. With regard to FCRA compliance, the violations cited most frequently were the failure to provide notices of adverse actions based on information contained in consumer reports or obtained from third parties.

Securities Regulators Oversee GLBA Compliance of Securities Firms

SEC, NASD, and NYSE Regulation oversee securities industry participants' compliance with GLBA's privacy and information safeguarding requirements. Similar to the banking agencies, they have issued rules and other guidance, conducted examinations of firms' compliance with federal securities laws and regulations, and, if appropriate, taken enforcement actions.

Regulations and Other Guidance

In June 2000, SEC adopted Regulation S-P, which implements GLBA's Title V information privacy and safeguarding requirements among the broker- dealers, investment companies, and SEC-registered investment advisers subject to SEC's jurisdiction. [77] Regulation S-P contains rules of general applicability that are substantively similar to the rules adopted by the banking agencies. In addition to providing general guidance, Regulation S-P contains numerous examples specific to the securities industry to provide more meaningful guidance to help firms implement its requirements. For example, the rule provides detailed guidance on the provision covering privacy and opt-out notices when a customer opens a brokerage account. It also contains a section regarding procedures to safeguard information, including the disposal of consumer report information. [78]

Since Regulation S-P was adopted, SEC staff have issued additional written guidance in the form of Staff Responses to Questions about Regulation S-P. According to SEC staff, companies also receive feedback on Regulation S-P compliance during the examination process, as well as during telephone inquiries made to SEC offices. However, unlike the federal banking agencies, SEC has issued no additional written guidance on institutions notifying customers in the event of unauthorized access to customer information. SEC staff said they are considering possible measures that would address information security programs in more detail, including the issue of how to respond to security breaches.

Examinations and Enforcement Actions

SEC has examined registered firms for Regulation S-P compliance. SEC staff said compliance with Regulation S-P was a focus area in SEC examinations during the first 1 to 11/2 years after July 2001, when it became effective. During this period, Regulation S-P compliance was reviewed in 858 broker-dealer examinations, of which 105 resulted in findings. [79] Also, during this period, Regulation S-P compliance was reviewed in 1,174 investment adviser examinations, of which 128 resulted in findings, and 218 investment company examinations, of which 17 resulted in findings.

SEC staff said that more recently SEC has adopted a risk-based approach to determine the depth of a review of compliance with Regulation S-P. Under this approach, an initial review of compliance with Regulation S-P is done to determine if a closer look is warranted. During the past 21/2 years, compliance with Regulation S-P was reviewed in 1,891 investment adviser examinations, of which 301 resulted in findings, and 257 investment company examinations, of which 20 resulted in findings. SEC staff said they had not broken out separate Regulation S-P examination findings of broker-dealer examinations for this period and could not provide those numbers. They said the most common deficiencies were failure to provide privacy notices, no or inadequate privacy policy, and no or inadequate policies and procedures for safeguarding customer information. SEC staff said they had not found any deficiencies during their exams that warranted formal enforcement actions. They told us they have dealt with Regulation S-P compliance more as a supervisory matter and required registrants to resolve deficiencies without taking formal actions.

SEC staff also said that SEC is now conducting a special review coordinated with NYSE Regulation looking at how broker-dealers are outsourcing certain functions that involve customer information. They said they are concerned with how registrants are managing the outsourcing process, including, among other things, due diligence in contractor selection, monitoring contractor performance, and disaster recovery/business continuity planning.

NASD and NYSE Regulation Oversee Compliance of Member Broker-Dealers

NASD and NYSE Regulation also oversee Regulation S-P compliance among member broker-dealers. According to NASD officials, NASD took a two-pronged approach to ensure that its members understand their obligations under Regulation S-P and comply with its requirements. First, NASD issued guidance to its members regarding requirements of the regulation. For example, when Regulation S-P was adopted, NASD issued guidance to facilitate compliance by providing a notice designed to inform and educate its members about Regulation S-P. [80] In the summer of 2001, NASD issued an article setting forth questions and answers regarding Regulation S-P and reminding members of the mandatory compliance deadline. [81] In July 2005, NASD issued another notice reminding members of their obligations relating to the protection of customer information. [82] Second, according to NASD officials, NASD conducts routine examinations—approximately 2,500 per year—to check compliance with NASD rules and the federal securities laws, including Regulation S-P. Examiners check compliance with Regulation S-P using a risk-based approach in which examiners review certain information such as supervisory review procedures to assess the controls that exist at a firm. Depending on its findings, NASD determines whether to inspect in more detail the firm's Regulation S-P policies and procedures to ensure they are reasonably designed to achieve compliance with Regulation S-P, including its safeguarding and privacy requirements. Regulation S-P compliance was reviewed in 4,760 NASD examinations of broker-dealers between October 1, 2000, and September 30, 2005. These examinations resulted in 502 informal actions and two formal actions—called Letters of Acceptance, Waiver, and Consent—for Regulation S-P violations. According to NASD, in one formal action, it censured and fined the respondents a total of $250,000 for various violations related to their failure to establish supervisory procedures and devote sufficient resources to supervision, including Regulation S-P compliance. In the other action, according to NASD, it censured and fined the firm and a principal associated person $28,500 and suspended the person for 30 days for failing to provide privacy notices to its customers and for several other non-privacy-related violations.

Similarly, NYSE Regulation issued guidance on Regulation S-P to its member firms and sent its members an information memo reminding them of Regulation S-P requirements shortly before they became mandatory. [83] NYSE Regulation's Sales Practice Review Unit conducts examinations of member firms' compliance with Regulation S-P and other privacy requirements on a 1-, 2- or 4-year cycle, or when the member firm is otherwise deemed to be at a certain level of risk.

State Insurance Regulators Require Insurers to Comply with Information Privacy and Security Provisions, but Enforcement May Be Limited

GLBA designates state insurance regulators as the authorities responsible for enforcement of its information privacy and safeguarding provisions among insurance companies. The individual states are responsible for enforcing GLBA with respect to insurance companies licensed in the state, and they may issue regulations. [84] The National Association of Insurance Commissioners (NAIC) has issued model rules to guide states in developing programs to enforce GLBA requirements and has sponsored a multistate review of insurance companies' performance in this regard.

NAIC Has Developed Model GLBA Privacy and Safeguarding Rules, but Not All States Have Adopted GLBA Regulations

NAIC has developed two model rules for states to use in developing regulations or laws to implement the GLBA information privacy and safeguarding provisions among the insurance companies they regulate. The first model rule, the Privacy of Consumer Financial and Health Information Regulation, issued in 2000, includes notice and opt-out requirements relating to insurance entities, and can be used by states as models for state laws and regulations. An August 2005 NAIC analysis showed that all states and the District of Columbia had adopted insurance laws or regulations to implement GLBA's requirements related to the privacy of financial information [85].

The second model rule, the Standards for Safeguarding Customer Information Model Regulation, issued in 2002, establishes standards for developing and implementing administrative, technical, and physical safeguards to protect the security, confidentiality, and integrity of customer information. In contrast to the privacy model, an October 2005 NAIC analysis showed that 17 states had yet to adopt a law or regulation setting standards for safeguarding customer information. In April 2002, GAO reported that insurance customer information and records in states that had not established safeguards may not be subject to a consistent level of legal protection envisioned by GLBA's privacy provisions [86].

Individual State Insurance Regulators Have Not Consistently Examined for Privacy and Security Compliance

Individual state insurance regulators have procedures for examining companies for compliance with information privacy and safeguarding requirements, but do not routinely do so. According to an NAIC official, NAIC's Market Conduct Examiners Handbook contains detailed examination procedures for reviewing information privacy requirements and its Financial Examiners Handbook has a segment devoted to security of computer-based systems. He said the individual state regulators can examine for compliance with privacy requirements as part of their comprehensive examinations of companies, but that states are focusing less on conducting comprehensive examinations and more on targeted examinations. As a result of a lack of complaints regarding privacy matters, however, he said the states are probably doing few targeted examinations of compliance with privacy requirements.

To forestall possible multiple, overlapping, and inconsistent examinations by numerous states, NAIC in 2005 sponsored a multistate review to gather information on insurance companies' compliance with GLBA privacy and safeguarding provisions. The review team, led by the District of Columbia's Department of Insurance, Securities and Banking (DISB), with the participation of 19 states, covered more than 100 of the largest insurance groups, representing about 800 insurance companies operating in the United States. [87] The review team administered a survey questionnaire, reviewed each insurer's responses to the questionnaire, and subsequently held conferences with representatives of the insurer. The review resulted in

- 22 findings related to the risk assessment process, including failure to work toward a formalized assessment process to identify risks of internal and external threats and hazards to the safeguarding, confidentiality, and integrity of information;
- 18 findings related to GLBA's requirements for information storage, transmission, and integrity;
- 16 findings related to the delivery of privacy notices (although 12 of those findings related to the provision of the initial notice rather than recurring findings); and
- no findings related to GLBA procedures for providing opt-out notifications or procedures for collecting opt-out elections.

These findings were similar to those of other financial regulators' examinations of GLBA compliance. However, unlike the other regulators, state insurance regulators do not have comparable examination programs to follow up to ensure that such findings are corrected and do not become more numerous. The DISB qualified the scope of its survey by noting that it did not include (1) a review of the insurer's efforts with respect to remediation activities, (2) a detailed analysis of the effectiveness of the insurer's plans to correct privacy problems or to protect the business against the consequences associated with any privacy-related occurrences, or (3) a determination of steps the insurer must take to become privacy compliant or maintain privacy compliance.

Although this survey was not a substitute for regulatory examination of insurers' compliance with GLBA, it could serve as a basis for further examination of such compliance. Other financial regulators have gathered preliminary information that they then use as a basis for further examinations of regulated entities. For example, in 2003, SEC followed up on reports of abusive practices in mutual fund trading by requesting information from various mutual fund companies on these trading practices, and this served as a basis for further examinations of individual companies. According to NAIC officials, the DISB survey results were never reviewed by state insurance regulators as part of their examinations of insurance companies. NAIC officials said the survey results were reviewed by NAIC's Market Analysis Working Group and referred back to DISB to determine what, if any, additional follow-up was necessary. DISB staff told us that most state insurance regulators, as well as DISB, do not have staff with adequate expertise to actually examine insurers' information privacy and safeguarding programs. They said the states would have to contract with vendors to obtain this expertise.

FTC Enforces GLBA and FCRA Compliance of Financial Institutions within Its Jurisdiction

As discussed earlier, FTC enforces GLBA for financial institutions not otherwise assigned to the enforcement authority of another regulator, and enforces FCRA for the same entities and others, including securities firms and insurance companies. FTC has issued rules implementing GLBA and FCRA information privacy and safeguarding requirements and developed other materials that provide detailed guidance for companies to implement the requirements. FTC issued two rules—referred to as the Privacy Rule and the Safeguards Rule—to implement GLBA's requirements for financial institutions not covered by similar regulations issued by the financial institution regulators. These rules provide examples to clarify things such as what constitutes a customer relationship and what types of information are covered under the law's sharing restrictions. FTC has also issued rules to implement the FACT Act amendments to FCRA, although some rules have not yet been issued in final form. [88] FTC provides additional guidance to financial institutions on how to comply with GLBA and FCRA in the form of business alerts, fact sheets, frequently asked questions, and a compliance guide for small businesses. For example, FTC has issued alerts on safeguarding customers' personal information, disposing of consumer report information, and insurers' use of consumer reports.

Between 2003 and 2005, FTC took enforcement actions against at least seven financial service providers for violations of GLBA information privacy and safeguarding requirements, resulting in settlement agreements with

- an Internet mortgage lender accused of false advertising and failure to protect sensitive consumer information;
- a credit card telemarketer that allegedly failed to notify consumers of its privacy practices and obtained information from consumers under false pretenses;
- two or more mortgage lenders charged with failing to protect consumers' personal information; and

- three nonprofit debt management organizations accused of failing to notify consumers how their personal information would be used, and other violations. [89]

NCUA, Securities, and Insurance Regulators Do Not Have Full Authority to Examine Third-Party Vendors, Including Information Resellers

As part of their bank examinations, FRB, FDIC, OCC, and OTS have authority to examine third-party service providers, such as some information resellers with which banks may do business. [90] Technology service provider examinations are done under the auspices of FFIEC and coordinated with other regulators. [91] Some vendors may be examined routinely; for example, officials of one information reseller providing services to banks told us that it is subject to periodic examinations under the auspices of FFIEC. In other cases, a service provider may be examined only once for a particular purpose. For example, OCC and FDIC examiners visited Acxiom, which provides a number of banks with information services, such as analyzing and enhancing customer information for marketing purposes. The examiners' visit focused on a security breach in which a client was granted access to information files obtained from other clients. According to Acxiom officials, this was a one-time review of the breach that occurred in its computer services operations and did not result in the company being added to a list of technology service providers that banking regulators routinely review.

Unlike the banking regulators, NCUA does not have authority to examine the third-party service providers of credit unions, including information resellers. [92] In 2003, we reported that credit unions increasingly rely on third-party vendors to support technology-related functions such as Internet banking, transaction processing, and fund transfers. [93] With greater reliance on third-party vendors, credit unions subject themselves to operational and reputational risks if they do not manage these vendors appropriately. While NCUA has issued guidance regarding the due diligence credit unions should apply to third-party vendors, the agency has no enforcement powers to ensure full and accurate disclosure. As such, in 2003 we suggested that Congress consider providing NCUA with legislative authority to examine third-party vendors, and NCUA has also requested such authority from Congress. However, an NCUA official told us that few of these vendors are information resellers because credit unions typically do not use them to a great extent. He said that credit unions generally use methods other than resellers to comply with PATRIOT Act customer identification requirements, and credit unions' bylaws typically forbid sharing customers' personal financial information for marketing purposes.

Similarly, federal securities regulators and representatives of state insurance regulators told us they generally do not have authority to examine or review the third-party service providers of the firms they oversee, including information resellers. According to SEC staff, the agency can examine the third-party vendor only if the firm also is an SEC-registered entity over which the agency has examination authority. However, they said that, to date, SEC has not seen sufficient problems with third-party vendors to justify requesting the authority to examine them at this time. They noted that in their examinations, they hold entities accountable for ensuring that personal information is appropriately safeguarded whether the information is managed in-house or by a vendor. Similarly, NASD officials said that although they do not have jurisdiction to oversee third-party vendors, their examiners review member firms' procedures for monitoring contractors, including whether such contracts contain clauses ensuring the privacy and security of customer information. In July 2005, NASD issued a Notice to Members reminding them that when they outsource certain activities as part of their business structure, they must conduct a due diligence analysis to ensure that the third-party service provider can adequately perform the outsourced functions and comply with federal securities laws and NASD rules. [94] Similarly, NYSE Regulation examinations review third-party contracts to ensure that they contain confidentiality clauses prohibiting the contractor from using or disclosing customer information for any use other than the purposes for which the information was provided to the contractor. NYSE Regulation has proposed a rule governing its members' use of contractors, which, if adopted, will require member firms to follow certain steps in selecting and overseeing contractors, such as applying prescribed due diligence standards and the record-keeping requirements of the securities laws [95].

State insurance regulators generally do not have authority to examine information resellers and other third-party service providers. NAIC officials told us that state insurance regulators can only examine information resellers or other companies if they are registered as rating organizations—companies that collect and analyze statistical information to assist insurance companies in their rate-making process. For example, NAIC said state insurance regulators can examine ISO—one of the resellers included in our review—because it is registered with states as a rating organization.

CONCLUSIONS

Advances in information technology and the computerization of records have spawned the growth of information reseller businesses, which regularly collect, process, and sell personal information about nearly all Americans. The information maintained by resellers commonly includes sensitive personal information, such as purchasing habits, estimated incomes, and Social Security numbers. The expansion in the past few decades in the sale of personal information has raised concerns about both personal privacy and data security. Many consumers may not be aware how much of their personal information is maintained and how frequently it is disseminated. In addition, identity theft has emerged as a serious problem, and data security breaches have occurred at some major resellers. At the same time, however, information resellers also provide some important benefits to both individuals and businesses. Financial institutions rely heavily on these resellers for a variety of vital purposes, including credit reporting (which reduces the cost of credit), PATRIOT Act compliance, and fraud detection. As Congress weighs various legislative options, it will need to consider the appropriate balance between protecting consumers' privacy and security interests and the benefits conferred by the current regime that allows a relatively free flow of information between companies.

No federal law explicitly requires all information resellers to safeguard all of the sensitive personal information they may hold. As we have discussed, FCRA applies only to consumer information used or intended to be used to help determine eligibility, and GLBA's safeguarding requirements apply only to customer data held by GLBA-defined financial institutions. Much of the personal information maintained by information resellers that does not fall under FCRA or GLBA is not necessarily required by federal law to be safeguarded, even when the information is sensitive and subject to misuse by identity thieves. Given financial institutions' widespread reliance on information resellers to comply with legal requirements, detect fraud, and market their products, the possibility for misuse of this sensitive personal information is heightened. Requiring information resellers to safeguard all of the sensitive personal information they hold would help ensure that explicit data security requirements apply more comprehensively to a class of companies that maintains large amounts of such data. Further, although the scope of this chapter focused on information resellers, this work has made clear to us that a wide range of retailers and other entities also maintain sensitive personal information on consumers. As Congress considers requiring information resellers to better ensure that all of the sensitive personal information they maintain is safeguarded, it may also wish to consider the potential costs and benefits of expanding more broadly the class of entities explicitly required to safeguard sensitive personal information. Any new safeguarding requirements would likely be more effectively implemented and least burdensome if, as with FTC's Safeguards Rule, they provided sufficient flexibility to account for the widely varying size and nature of businesses that hold sensitive personal information.

The proliferation of sensitive personal information in the marketplace and increasing numbers of high-profile data breaches have motivated many states to enact data security laws with breach notification requirements. No federal statute currently requires breach notification, but such legislation could have certain benefits. Companies would have incentives to improve data safeguarding to reduce the reputational risk of a publicized breach, and consumers would know to take potential action against a risk of identity theft or other related harm. Congress has held many hearings related to data breaches, and several bills have been introduced that would require breach notification. We support congressional actions to require information resellers, and other companies, to notify individuals when breaches of sensitive information occur. In previous work, we have also identified key benefits and challenges of notifying the public about security breaches that occur at federal agencies. To be cost effective and reduce unnecessary burden on consumers, agencies, and industry, it would be important for Congress to identify a threshold for notification that would allow individuals to take steps to protect themselves where the risk of identity theft or other related harm exists, while ensuring they are only notified in cases where the level of risk warrants such action. Objective criteria for when notification is required and appropriate enforcement mechanisms are also important considerations. Congress should also consider whether and when a federal breach notification law would preempt state laws.

FTC has taken many significant enforcement actions against information resellers and other companies that have violated federal privacy laws, and it is important that the agency have the appropriate enforcement remedies. Unlike FCRA, GLBA does not provide FTC with civil penalty authority, and agency staff have expressed concerns that the remedies FTC has available under GLBA—such as disgorgement and consumer redress—are impractical enforcement tools for violations involving breaches of mass consumer data. Providing FTC with the authority to seek civil penalties for violations of GLBA could help the agency more effectively enforce that law's safeguarding provisions.

Federal financial regulators generally appear to provide suitable oversight of their regulated entities' compliance with privacy and information security laws governing consumer information. The regulators do not typically distinguish between data that entities receive from resellers and other sources, but this seems reasonable given that the sensitivity, rather than the source, of the data is the most important factor in examining data security practices. However, state insurance regulators do not have comparable examination programs to other financial regulators to ensure consistent GLBA compliance. This may be a source of concern given the recent multistate survey that identified deficiencies in GLBA compliance at insurance companies.

MATTERS FOR CONGRESSIONAL CONSIDERATION

Safeguarding provisions of FCRA and GLBA do not apply to all sensitive personal information held by information resellers. To ensure that such data are protected on a more consistent basis, Congress should consider requiring information resellers to safeguard all sensitive personal information they hold. As Congress considers how best to protect data maintained by information resellers, it should also consider whether to expand more broadly the class of entities explicitly required to safeguard sensitive personal information. If Congress were to choose to expand safeguarding requirements, it should consider providing the implementing agencies with sufficient flexibility to account for the wide range in the size and nature of entities that hold sensitive personal information.

To ensure that the Federal Trade Commission has the tools it needs to most effectively act against data privacy and security violations, Congress should consider providing the agency with civil penalty authority for its enforcement of the Gramm-Leach-Bliley Act's privacy and safeguarding provisions.

RECOMMENDATION FOR EXECUTIVE ACTION

We recommend that state insurance regulators, individually and in concert with the National Association of Insurance Commissioners, take additional measures to ensure appropriate enforcement of insurance companies' compliance with the privacy and safeguarding provisions of the GrammLeach-Bliley Act. As a first step, state insurance regulators and NAIC should follow up appropriately on deficiencies related to compliance with these provisions that were identified in the recent nationwide survey as part of a broader targeted examination of GLBA privacy and safeguarding requirements.

AGENCY COMMENTS

We provided a draft of this chapter to FDIC, FRB, FTC, NAIC, NASD, NCUA, NYSE Regulation, OCC, OTS, and SEC for comment. These agencies provided technical comments, which we incorporated, as appropriate. In addition, FTC provided a written response, which is reprinted in appendix III. In its response, FTC noted that it has previously recommended that Congress consider legislative actions to increase the protection afforded personal sensitive data, including extending GLBA safeguarding principles to other entities that maintain sensitive information. FTC also noted that it concurs with our finding that a civil penalty often is the most appropriate and effective remedy in cases under GLBA privacy and safeguarding provisions.

REFERENCES

[1]   This chapter uses "information resellers" to describe businesses that collect and resell personal information, but there is no one commonly agreed-upon term for such companies. FTC has sometimes used the term "data brokers" but the companies themselves typically use other terms, such as "information solutions providers."

[2]   The Fair Credit Reporting Act, Pub. L. No. 90-321, title VI (May 29, 1968) as added by Pub. L. No. 9 1-508, title VI, § 601, 84 Stat. 1128 (Oct. 26, 1970) (codified at 15 U.S.C. § 1681-168 1x); and Title V of the Gramm-Leach-Bliley Act (Financial Services Modernization Act of 1999), Pub. L. No. 106-102, title V, subtitle A, 113 Stat. 1338 (Nov. 12, 1999) (codified at 15 U.S.C. § 6801-6809). As discussed later in this chapter, other federal laws—such as the Driver's Privacy Protection Act of 1994 and the Health Insurance Portability and Accountability Act of 1996—also govern the use and sharing of certain types of personal information.

[3]   For more information about Internet resellers, see GAO, Social Security Numbers: Internet Resellers Provide Few Full SSNs, but Congress Should Consider Enacting Standards for Truncating SSNs, GAO-06-495 (Washington, D.C.: May 17, 2006).

[4]   We use "nationwide credit bureau" and "nationwide consumer reporting agency" interchangeably in this chapter, and they have the same meaning as the FCRA phrase "consumer reporting agency that compiles and maintains files on consumers on a nationwide basis." FCRA defines this phrase as a consumer reporting agency that regularly engages in the practice of assembling or evaluating, and maintaining public record information and credit account information for the purpose of furnishing consumer reports to third parties bearing on a consumer's credit worthiness, credit standing, or credit capacity. 15 U.S.C. § 1681a(p).

[5]   For information about federal agencies' use of information resellers, see GAO, Personal Information: Agency and Reseller Adherence to Key Privacy Principles, GAO-06-421 (Washington, D.C.: Apr. 4, 2006).

[6]   Credit header data are the nonfinancial identifying information located at the top of a credit report, such as name, current and prior addresses, telephone number, and Social Security number.

[7]   This chapter focuses on how financial institutions use data from information resellers in conducting transactions with consumers. We did not review other ways that financial institutions use information resellers, such as to screen their potential employees or to gather information about other businesses.

[8]   The three nationwide credit bureaus use software models developed by the Fair Isaac Corporation to produce FICO® credit scores, which are credit scores used by many financial services firms. In March 2006, the bureaus announced they will begin selling a new credit score that they developed jointly. The score will be calculated the same way for each credit bureau to enhance consistency among all three bureaus.

[9]   A nationwide specialty CRA is defined in FCRA to mean a CRA that compiles and maintains files on consumers on a nationwide basis relating to medical records or payments; residential or tenant history; check-writing history; employment history; or insurance claims. 15 U.S.C. § 168 1a(w).

[10] Uniting and Strengthening America by Providing Appropriate Tools Required to Intercept and Obstruct Terrorism (USA PATRIOT ACT) Act of 2001, Pub. L. No. 107-56, 115 Stat. 272 (Oct. 26, 2001). We will refer to the act as the PATRIOT Act.

[11] Title III of the PATRIOT Act (cited as the "International Money Laundering Abatement and Financial Anti-Terrorism Act of 2001") amended the U.S. government's anti-money laundering regulatory structure. For instance, section 326 added new requirements for the Secretary of the Treasury and the federal financial regulators to issue regulations setting forth minimum standards for financial institutions to (1) verify the identity of persons seeking to open an account; (2) maintain records of the information used to verify a person's identity, including name, address, and other identifying information; and (3) consult lists of known or suspected terrorists or terrorist organizations provided to the financial institution by any government agency to determine whether a person seeking to open an account appears on the list. See 31 U.S.C. § 5318(l). Section 326 requirements for customer verification apply to financial institutions broadly, including, among others, financial institutions that are subject to regulation by one of the federal banking regulators, as well as nonfederally insured credit unions, private banks and trust companies; securities broker-dealers; futures commission merchants and introducing brokers; and mutual funds. 31 U.S.C. § 5312 and 31 C.F.R. § Part 103.

[12] A manufacturer may request that consumers submit their contact information on a warranty card in the event of a product malfunction or insurance claim. For marketing purposes, many warranty cards request additional information on such things as the gender and age of household occupants, occupation and income information, spending habits, and lifestyle interests; this information is sometimes sold to information resellers.

[13] The Fair Credit Reporting Act, described in more detail below, generally permits prescreening only if the financial institution makes a firm offer of credit or insurance for all consumers who meet the criteria for the credit or insurance being offered. 15 U.S.C. § 1681b(c)(1)(B).

[14] This chapter focuses on the use and sharing of personal information among private sector entities, and therefore we only describe laws governing these entities. Other laws, primarily the Privacy Act of 1974, govern the collection and use of personal information by government agencies. See Pub. L. No. 93-579, 88 Stat. 1896 (Dec. 31, 1974), codified at 5 U.S.C. § 552a.

[15] The Health Insurance Portability and Accountability Act of 1996, Pub. L. No. 104-191, § 262, 110 Stat. 1936 (Aug. 21, 1996), codified at 42 U.S.C. §§ 1320d – 1320d-8, protects the privacy of individually identifiable health information. The scope of this work did not include the collection and use of health information.

[16] Pub. L. No. 103-322, title XXX, 108 Stat. 2099 (Sept. 13, 1994) (codified at 18 U.S.C. §§ 2721 - 2725).

[17] 18 U.S.C. § 2721(b)(11).

[18] Pub. L. No. 63-203, ch. 311, 38 Stat. 717 (Sept. 26, 1914) (codified at 15 U.S.C. §§ 41 – 58).

[19] See 12 U.S.C. § 1867 (FRB, FDIC, and OCC); and 12 U.S.C. § 1464(d)(7) (OTS).

[20] Although the scope of this chapter is limited to federal privacy and data security laws, many states have laws of their own that apply to the activities of information resellers.

Many of these laws require companies to notify consumers when their personal data may have been lost or stolen. For example, in 2002, California enacted a database breach notification act (Cal. Civ. Code § 1798.82), which requires disclosure of any security breach of data to any state resident whose unencrypted personal information was, or is reasonably believed to have been, acquired by an unauthorized person.

[21] FCRA defines a "consumer report" as "any written, oral, or other communication of any information by a consumer reporting agency bearing on a consumer's credit worthiness, credit standing, credit capacity, character, general reputation, personal characteristics, or mode of living which is used or expected to be used or collected in whole or in part for the purpose of serving as a factor in establishing the consumer's eligibility for (A) credit or insurance to be used primarily for personal, family, or household purposes; (B) employment purposes; or (C) any other purpose authorized under [15 U.S.C. § 1681b]." 15 U.S.C. § 1681a(d)(1).

[22] Pub. L. No. 108-159, 117 Stat. 1952 (Dec. 4, 2003) (codified at 15 U.S.C. §§ 1681c-1, 1681c2, 1681x, 1681s-3, 1681w).

[23] We did not determine which information reseller databases are subject to FCRA. The information we include is based on what information resellers told us about how FCRA applies to their activities.

[24] Consumers also have the right to receive a free copy of their credit file from CRAs when they have been victims of identity theft or are subject to an adverse action as a result of information in their file, or in certain other circumstances where they are unemployed, recipients of public welfare, or have reason to believe that their file contains inaccurate information due to fraud.

[25] FCRA also provides certain other opt-out rights concerning affiliate sharing. See 15 U.S.C. §§ 168 1a(d)(2)(iii); and 168 1s-3. In addition to FCRA, GLBA requires that financial institutions allow their customers to opt out of the sharing of their nonpublic personal information with nonaffiliated companies, unless the sharing falls under an exception under GLBA. See 15 U.S.C. § 6802.

[26] 16 C.F.R. § 610.2.

[27] 16 C.F.R. § 610.3.

[28] 15 U.S.C. § 6802.

[29] See 15 U.S.C. § 6809(9). GLBA defines a consumer as "an individual who obtains, from a financial institution, financial products or services which are to be used primarily for personal, family, or household purposes." Thus, GLBA does not apply to a business customer, such as a sole proprietor. 16 C.F.R. § 313.3(e). A "customer" means a consumer who has a "customer relationship"—that is, a continuing relationship with the financial institution.

[30] 15 U.S.C. § 6802(e)(3)(B) and (6).

[31] 15 U.S.C. § 6802(e)(1)(A).

[32] 15 U.S.C. § 6809(3)(A).

[33] 12 U.S.C. § 1843(k). This is a list of nonbanking activities determined by FRB as of the date of GLBA's enactment to be "so closely related to banking or managing or controlling banks as to be a proper incident thereto." See 12 C.F.R. § 225.28 (1999). FDIC, FRB, NCUA, OCC, OTS and SEC in their implementing GLBA regulations define the term "financial institution" as those institutions in the business of engaging in activities that are financial in nature or incidental to such financial activities. See 12

C.F.R. §§ 40.3(k)(1) (OCC), 216.3(k)(1) (FRB), 332.3(k)(1) (FDIC), 573.3(k)(1) (OTS), and 716.3(l)(1) (NCUA); and 17 C.F.R. § 248.3(n)(1) (SEC). See 16 C.F.R. § 313.3(k)(1) (FTC).

[34] 16 C.F.R. § 313. 18(a)(2); and 65 Fed. Reg. 33646, 33654 (May 24, 2000).

[35] 16 C.F.R. §§ 313.3(k)(1) and (3)(iv).

[36] 12 C.F.R. § 225.28(b)(2)(v) (1999). FRB described credit bureau services as those services "maintaining information related to the credit history of consumers and providing the information to a credit grantor who is considering a borrower's application for credit or who has extended credit to the borrower."

[37] See Trans Union LLC v. FTC, 295 F.3d 42, 48 (D.C. Cir. 2002); and 16 C.F.R. § 3 13.3(k).

[38] A representative of the company noted that, as required by law, the data used for these two products are kept in separate databases that are not commingled.

[39] 16 C.F.R. § 313.11 (FTC); see also 12 C.F.R. §§ 40.11 (OCC), 216.11 (FRB), 332.11 (FDIC), 573.11 (OTS), and 716.11 (NCUA); and 17 C.F.R. § 248.11 (SEC). The regulations were upheld in Individual Reference Services Group, Inc. v. FTC, 145 F. Supp.2d 6, 34 – 35 (D. DC 2002) ("the use restrictions affirmatively imposed by the Regulations are consistent with the purpose of the GLB Act").

[40] The FTC regulation states: "[y]ou may disclose and use the information pursuant to [a GLBA exception] in the ordinary course of business to carry out the activity covered by the exception under which you received the information." 16 C.F.R. § 313.11(a)(1)(iii).

[41] See 15 U.S.C. § 6802(c), which states: "[A] nonaffiliated third party that receives from a financial institution nonpublic personal information . . . shall not . . . disclose such information to any other person that is a nonaffiliated third party of both the financial institution and such receiving third party, unless such disclosure would be lawful if made directly to such other person by the financial institution." This provision is commonly referred to as GLBA's reuse and redisclosure provision. See 16 C.F.R. § 313.11(b)(1)(iii).

[42] See 15 U.S.C. § 6801 note.

[43] The company said that it does not allow information collected for its FCRA-regulated database to be used to update the "pre-GLBA" database.

[44] 15 U.S.C. § 6801.

[45] See, for example, 16 C.F.R. § 314.3 (FTC).

[46] See, for example, 16 C.F.R. § 314.4(d).

[47] The settlement will require BJ's Wholesale Club to implement a comprehensive information security program and obtain audits by an independent third-party security professional every other year for 20 years. In the Matter of BJ's Wholesale Club, Inc., F.T.C. No. 0423160 (2005). A consent agreement does not constitute an admission of a violation of law.

[48] Prepared Statement of the Federal Trade Commission on "Data Breaches and Identity Theft" Before the Senate Comm. on Commerce, Science, and Transportation, 109th Cong., 1st Sess. (2005).

[49] Although there is no applicable federal statute governing notification of data breaches, the banking agencies have issued guidance to financial institutions under their jurisdiction requiring them in some cases to notify customers affected by a data breach. States that have enacted breach notification requirements include Arizona, Arkansas,

California, Colorado, Connecticut, Delaware, Florida, Georgia, Hawaii, Idaho, Illinois, Indiana, Kansas, Louisiana, Maine, Minnesota, Montana, Nebraska, Nevada, New Jersey, New York, North Carolina, North Dakota, Ohio, Pennsylvania, Rhode Island, Tennessee, Texas, Utah, Vermont, Washington, and Wisconsin. Many other states have introduced legislation.

[50] United States v. ChoicePoint, Inc., No. 1:06-cv-00198-JTC (N.D. Ga., Feb. 15, 2006). As part of the settlement, ChoicePoint admitted no violations of law. According to ChoicePoint, the company has taken steps since the breach to enhance its customer screening process and to assist affected consumers.

[51] Congressional Research Service, Personal Data Security Breaches: Context and Incident Summaries, Order Code RL33199 (Washington, D.C., Dec. 16, 2005).

[52] For example, Identity Theft: Recent Developments Involving the Security of Sensitive Consumer Information: Hearing Before the Senate Comm. on Banking, Housing, and Urban Affairs, 109th Cong., 1st Sess. (2005); Securing Electronic Personal Data: Striking a Balance Between Privacy and Commercial and Governmental Use: Hearing Before the Senate Comm. on the Judiciary, 109th Cong., 1st Sess. (2005); Assessing Data Security: Preventing Breaches and Protecting Sensitive Information: Hearing Before the House Comm. on Financial Services, 109th Cong., 1st Sess. (2005); Securing Consumers' Data: Options Following Security Breaches: Hearing Before the Subcomm. On Commerce, Trade, and Consumer Protection of the House Comm. on Energy and Commerce, 109th Cong., 1st Sess. (2005).

[53] For more information on the key benefits and challenges associated with notifying the public about security breaches, see GAO, Privacy: Preventing and Responding to Improper Disclosures of Personal Information, GAO-06-833T (Washington, D.C.: June 8, 2006).

[54] FCRA gives enforcement authority to FDIC, FRB, OCC, OTS, and NCUA over their banks, thrifts, and credit unions, among other entities. FCRA assigned regulatory authority to the Departments of Transportation and Agriculture over entities under their jurisdiction. 15 U.S.C. § 1681s.

[55] 15 U.S.C. § 6805. GLBA required FTC and other regulators with responsibilities under the statute to issue consistent and comparable regulations. 15 U.S.C. § 6804.

[56] 15 U.S.C. § 168 1s(c).

[57] Conn. Gen. Stat. Anno. §§ 36a-41 - 44 (disclosure to broker-dealers or investment advisers engaged in contractual networking arrangements with the financial institution permitted after the customer is given notice and an opportunity to opt out); N.D. Cent. Code §§ 6.08.1- 01 - 10; Vt. Stat. Anno. Tit 8, §§ 10201 – 10205.

[58] For instance, FTC staff told us the agency filed suit in the following cases: In the Matter of Credit Bureau of Lorain, Inc., 81 F.T.C. 381 (1972); In the Matter of Credit Bureau of Columbus, Inc., 81 F.T.C. 938 (1972); In the Matter of Credit Bureau of Greater Syracuse, Inc., 84 F.T.C. 1660 (1974); In the Matter of Robert N. Barnes, 85 F.T.C. 520 (1975); In the Matter of Filmdex Chex System, Inc., 85 F.T.C. 889 (1975); In the Matter of Credit Data Northwest, 86 F.T.C. 389 (1975); In the Matter of Interstate Check Systems, Inc., 88 F.T.C. 984 (1976); In the Matter of Moore and Associates, Inc., 92 F.T.C. 440 (1978); In the Matter of Howard Enterprises, 93 F.T.C. 909 (1979); In the Matter of Trans Union Credit Information Co., 102 F.T.C. 1109 (1983); FTC v. TRW Inc., 784 F. Supp. 361 (N.D. Tex. 1991); In the Matter of I.R.S.C., Inc.,

116 F.T.C. 266 (1993); In the Matter of CDB Infotek, 116 F.T.C. 280 (1993); In the Matter of Inter-Fact Inc., 116 F.T.C. 294 (1993); In the Matter of W.D.I.A.Corp., 117 F.T.C. 757 (1994); In the Matter of Equifax Credit Information Services, Inc., 120 F.T.C. 577 (1995). See also United States v. ChoicePoint, Inc., No. 1:06- cv-00198-JTC (N.D. Ga. Feb. 15, 2006); United States v. Far West Credit, Inc., No. 2:06-cv00041-TC (C.D. Utah Jan. 17, 2006); and In the Matter of Southern Maryland Credit Bureau, Inc., 101 F.T.C. 19 (1983).

[59] In 1996, TRW Inc. sold its credit reporting business to a group of investors, who named the new company Experian.

[60] FTC has also enforced FCRA against resellers for other types of violations. For example, in 2000 FTC settled with the three nationwide credit bureaus after alleging that consumers were unable to adequately access the companies' personnel by telephone to discuss or dispute possible errors in their files. United States v. Equifax Credit Information Services, Inc., No. 1:00-CV-0087 (N.D. Ga. 2000); United States v. Experian Information Solutions, Inc., 3-00CV0056-L. (N.D. Tx. 2000); and United States v. Trans Union LLC, No. 00C 0235 (N.D. Ill. 2000). See http://www.ftc.gov/opa/2000/01/busysignal.htm. A consent agreement does not constitute an admission of a violation of law.

[61] In the Matter of Equifax Credit Information Services, Inc., 120 F.T.C. 577 (1995). A consent agreement does not constitute an admission of a violation of law.

[62] In the Matter of Trans Union Corp., F.T.C. No. 9255, 2000 WL 257766 (2000), petition for review denied, 245 F.3d 809 (D.C. Cir. 2001).

[63] United States v. ChoicePoint, Inc., No. 1:06-cv-00198-JTC (N.D. Ga., Feb. 15, 2006).

[64] Injunctions are judicial orders commanding a party to take an action or prohibiting a party from doing or continuing to do a certain activity. Disgorgement is having to give up profits or other gains illegally obtained.

[65] 15 U.S.C. § 1681s and 15 U.S.C. § 45(l) and (m). Regarding GLBA's prohibition against fraudulent access to financial information where a person obtains financial information relating to another person under false pretenses (pretext provisions), GLBA allows FTC to seek civil penalties for violations. Specifically, FTC has authority to enforce the GLBA pretext provisions in the same manner and with the same power and authority as it has under the Fair Debt Collection Practices Act (codified at 15 U.S.C. §§ 1692 – 1692o). 15 U.S.C. § 6822(a). A violation of the Fair Debt Collection Practices Act is deemed by federal law to be an unfair or deceptive act or practice in violation of the FTC Act, which means that FTC may impose civil penalties. 15 U.S.C. § 1692l(a); and United States v. National Financial Services, Inc., 98 F.3d 131, 139 - 141 (4th Cir. 1996). According to FTC officials, they do not have similar civil penalty authority for violations of GLBA's privacy and safeguarding provisions.

[66] 12 U.S.C. § 1818(i)(2)(A)(i).

[67] Some exceptions may exist. For example, section 411 of the FACT Act (which amended section 604(g) of FCRA (12 U.S.C. 1681b(g))), generally limits with certain exceptions creditors' ability to obtain or use medical information pertaining to a consumer for credit purposes. This section requires the banking regulatory agencies and NCUA to issue regulations relating to the use of medical information in credit transactions. The regulations apply broadly, and the exceptions therein are available to

all creditors, not just the financial institutions supervised by those agencies. See final rule published at 70 Fed. Reg. 70664, 70665 - 6 (Nov. 22, 2005).

[68] In addition to the responsibilities assigned to financial institution regulators and FTC, FCRA assigns enforcement authority to the Departments of Transportation and Agriculture for entities subject to their oversight, such as transportation carriers.

[69] The various banking agency GLBA and FCRA regulations can be found at 12 C.F.R. Parts 40 and 41 (OCC); 12 C.F.R. Parts 216, 222, and 232 (FRB); 12 C.F.R. Parts 332 and 334 (FDIC); 12 C.F.R. Parts 573 and 571 (OTS); and 12 C.F.R. Parts 716 and 717 (NCUA).

[70] 65 Fed. Reg. 35162 (June 1, 2000); and 65 Fed. Reg. 31722 (May 18, 2000). OCC, FRB, OTS, and FDIC issued their rules jointly. All of the rules were substantively identical but contained differences to account for differences between the agencies' legal authorities and, as appropriate, for the types of institutions within each agency's jurisdiction.

[71] 66 Fed. Reg. 8616 (Feb. 1, 2001) ("Interagency Guidelines Establishing Standards for Safeguarding Customer Information") (renamed "Interagency Guidelines Establishing Information Security Standards," 70 Fed. Reg. 15736 (Mar. 29, 2005)).

[72] 70 Fed. Reg. 15736 (Mar. 29, 2005) ("Interagency Guidance on Response Programs for Unauthorized Access to Customer Information and Customer Notice").

[73] Pub. L. No. 108-109, 117 Stat. 1952 (Dec. 4, 2003).

[74] See 15 U.S.C. § 1681w; 69 Fed. Reg. 77610 (Dec. 28, 2004); and 69 Fed. Reg. 68690 (Nov. 24, 2004).

[75] The examinations are risk-based and conducted in cycles depending on the institution's condition and size. Banking regulators are required by law, 12 U.S.C. § 1820(d), to examine insured institutions for safety and soundness at least once during each 12-month period, except for smaller institutions that meet specified conditions that can be examined each 18-month period. We use the term "thrifts" to refer to savings associations.

[76] Banking regulators have broad enforcement powers and can take formal actions (cease and desist orders, civil money penalties, removal orders, and suspension orders, among others) or informal enforcement actions (such as memoranda of understanding and board resolutions). Informal actions are generally not publicly disclosed.

[77] 65 Fed. Reg. 40334 (June 29, 2000), codified at 17 C.F.R. Part 248. SEC, NASD, and NYSE Regulation regulate broker-dealers by, among other things, examining their operations and reviewing customer complaints. SEC evaluates the quality of NASD and NYSE oversight in enforcing their members' compliance with federal securities laws through self-regulatory organization oversight inspections and broker-dealer oversight examinations. SEC is the primary regulator of investment companies and investment advisers registered with the SEC.

[78] 17 C.F.R. § 248.30.

[79] An examination finding would be any compliance deficiency (including an internal control weakness) or violation requiring corrective action.

[80] NASD Notice to Members 00-66 (September 2000).

[81] NASDR Regulatory and Compliance Alert (Summer 2001).

[82] NASD Notice to Members 05-49 (July 2005).

[83] NYSE Information Memoranda Nos. 01-10 (June 19, 2001) and 01-13 (June 21, 2001).

[84] 15 U.S.C. § 6805(a)(6). State insurance authorities may enforce GLBA and may establish privacy regulations. However, GLBA mandates that state insurance authorities establish standards for safeguarding customer information and that the standards be implemented by rules. 15 U.S.C. §§ 6801(b) and 6805(b)(2). Moreover, if a state insurance authority fails to adopt regulations to carry out GLBA's privacy and safeguarding provisions, the state forfeits its eligibility under GLBA to override certain customer protection regulations promulgated by the federal depository institution regulators applicable to insurance sales by or at depository institutions. 15 U.S.C. § 6805(c).

[85] We did not corroborate or independently verify NAIC's analysis.

[86] GAO, Financial Privacy: Status of State Actions on Gramm-Leach-Bliley Act's Privacy Provisions, GAO-02-361 (Washington, D.C.: Apr. 12, 2002).

[87] District of Columbia, Department of Insurance, Securities and Banking, Preliminary Report: Status of Insurance Industry Practices and Procedures to Protect the Privacy of Customer Information (September 2005). According to department staff, the final report is pending. The staff said the preliminary and final results should not differ because the preliminary results included responses of more than 90 percent of the companies, including all of the large companies.

[88] FTC's GLBA and FCRA regulations can be found at 16 C.F.R. Parts 313 and 314 and 16 C.F.R. Parts 600 through 698.

[89] FTC v. 30 Minute Mortgage, Inc., No. 03-60021-CIV (S.D. Fla. 2003); FTC v. Sainz Enterprises LLC, No. 04WM-2078 (CBS) (D. Co. 2004); In the Matter of Superior Mortgage Corp., F.T.C. No. 052- 3136 (2005); In the Matter of Sunbelt Lending Servs., FTC No. C4129 (2005); In the Matter of Nationwide Mortgage Group, Inc., F.T.C. No 9319 (2005); FTC v. Nat'l. Consumer Council, Inc., No. SACV04-0474CJC (JWJX) (C.D. Cal. 2005); FTC v. Debt Mgmt. Found. Serv., Inc., No. 8:04-cv-01674-EAK-MSS (M.D. Fla. 2005). A consent agreement does not constitute an admission of a violation of law.

[90] See 12 U.S.C. § 1867 (FRB, FDIC, and OCC); and 12 U.S.C. § 1464(d)(7) (OTS).

[91] In January 2006, we reported on contractors' access to and sharing of Social Security numbers and federal oversight of regulated entities that contract for services. See GAO, Social Security Numbers: Stronger Protections Needed When Contractors Have Access to SSNs, GAO-06-238 (Washington, D.C.: Jan. 23, 2006).

[92] NCUA had temporary authority to examine third-party service providers under the Examination Parity and Year 2000 (Y2K) Readiness for Financial Institutions Act, Pub. L. No. 105-164, 112 Stat. 32 (Mar. 20, 1998) but that authority expired as of December 31, 2001. 12 U.S.C. § 1786a(c) and (f).

[93] GAO, Credit Unions: Financial Condition Has Improved, but Opportunities Exist to Enhance Oversight and Share Insurance Management, GAO-04-91 (Washington, D.C.: Oct. 27, 2003).

[94] NASD Notice to Members 05-48 (July 2005).

[95] SR-NYSE-2005-22, Proposed Rule 340, Outsourcing: Due Diligence and Conditions in the Use of Service Providers, and Proposed Amendments to Rule 342, Offices - Approval, Supervision and Control (Mar. 16, 2005).

APPENDIX I: SCOPE AND METHODOLOGY

Our chapter objectives were to examine (1) how financial institutions use data products supplied by information resellers, the types of information contained in these products, and the sources of the information; (2) how federal laws governing the privacy and security of personal data apply to information resellers, and what rights and opportunities exist for individuals to view and correct data held by resellers; (3) how federal financial institution regulators and the Federal Trade Commission (FTC) oversee information resellers' compliance with federal privacy and information security laws; and (4) how federal financial institution regulators, state insurance regulators, and FTC oversee financial institutions' compliance with federal privacy and information security laws governing consumer information, including information supplied by information resellers.

For the purposes of this chapter, we defined "information resellers" broadly to refer to businesses that collect and aggregate personal information from multiple sources and make it available to their customers. The three nationwide credit bureaus were included in this definition. Our audit work focused primarily on larger information resellers and did not cover smaller Internet-based resellers because these companies were rarely or never used by financial institutions from which we collected information. Our scope was limited to resellers' use and sale of personal information about individuals; it did not include other information that resellers may provide, such as data on commercial enterprises. Our review of financial institutions covered the banking, securities, property and casualty insurance, and consumer lending and finance industries, but excluded life insurance and health insurance companies because they use health data that are covered by federal laws that were outside the scope of our work. In addition, we included financial institutions' use of reseller information for purposes related to customers and other consumers, but excluded their use of reseller products for screening their own employees or making business decisions such as where to locate a facility.

To address all of the objectives, we interviewed or received written responses from 10 information resellers—Acxiom, eFunds, ChoicePoint, Equifax, Experian, LexisNexis, ISO, Regulatory DataCorp, Thompson West, and TransUnion. We also reviewed marketing materials, sample contracts, sample reports, and other items from these companies that provided detailed information on the data contained in their products. These companies were selected because, according to the financial institutions, trade associations, and industry experts we spoke with, they constitute most of the largest and most significant information resellers offering services to the financial industry sector, and collectively they represent a variety of different products. The information resellers we included and the products they offer do not necessarily represent the full scope of the industry. We also spoke with representatives of the Consumer Data Industry Association and the Direct Marketing Association, trade associations that represent portions of the information reseller industry.

To determine how financial institutions use data products supplied by information resellers and the types and sources of the data, we also interviewed or received written responses, and collected and analyzed documents, from knowledgeable representatives at financial institutions in the banking, securities, property and casualty insurance, and consumer lending and finance industries. We gathered information from Bank of America, Citigroup, and JPMorgan Chase, which are the three largest U.S. bank holding companies by asset size, as well as Goldman Sachs, Morgan Stanley, and Merrill Lynch, which are the three largest global securities firms by revenue. We also interviewed representatives at American International Group, State Farm, and Allstate, which are the three largest U.S. insurance companies and include the two largest property/casualty insurers. We also interviewed representatives at GE Consumer Finance, one of the world's 10 largest consumer finance companies, and four other financial institutions—American Express, Wells Fargo Financial, Security Finance, and Check into Cash—which together offer a variety of consumer lending products, including automobile financing, credit cards, and payday loans. We also interviewed officials at trade associations representing these financial services industries, including the American Bankers Association, Independent Community Bankers of America, Securities Industry Association, Investment Company Institute, American Insurance Association, and American Financial Services Association.

These financial institutions from which we gathered information conduct a significant portion of the transactions in the financial services sector. For example, they collectively own 9 of the 50 largest commercial depository institutions, holding about 20 percent of total domestic deposits, as well as 8 of the 10 largest credit card issuers. The insurance companies we spoke with represent about a quarter of the U.S. property and casualty insurer market share. In most cases, we selected these financial institutions by determining the largest companies in each of the four industries, based on data from reputable sources. In two cases, we spoke with firms because they were recommended by representatives of their trade association. Our findings on how financial institutions use information resellers are not representative of the entire financial services industry. However, we believe they accurately represent institutions' use of resellers because our findings from discussions with these companies and their representatives were corroborated by discussions with information resellers, regulators, legal experts, and privacy and consumer advocacy groups.

To identify how federal privacy and data security laws and regulations apply to information resellers and individuals' rights and opportunities to view and correct reseller data, we reviewed and analyzed relevant federal laws, regulations, and guidance. We also met with staff of the Board of Governors of the Federal Reserve System, Federal Deposit Insurance Corporation, Federal Trade Commission, National Credit Union Administration, Office of the Comptroller of the Currency (OCC), Office of Thrift Supervision, and Securities and Exchange Commission, as well as the National Association of Insurance Commissioners (NAIC), NASD (formerly known as the National Association of Securities Dealers), New York Stock Exchange Regulation (NYSE Regulation), and the District of Columbia's Department of Insurance, Securities and Banking (DISB). In addition, we interviewed three legal experts in the area of privacy law that work in academia or represent financial institutions and information resellers. We also interviewed and collected documents from information resellers, financial institutions, federal regulators, and a variety of privacy and consumer advocacy groups, to gather views on the applicability of laws to information resellers and the adequacy of existing laws.

To describe how regulators oversee information resellers' and financial institutions' compliance with federal privacy and data security laws, we met with the federal agencies, financial institutions, information resellers, and other parties listed above. We also reviewed federal agencies' guidance, examination procedures, settlement agreements, and other documents, as well as relevant reports and documents from NAIC, NASD, and NYSE Regulation. To help illustrate regulators' examination activities in this area, we also met with

OCC staff who conduct examinations at three national banks and reviewed their examination workpapers. We also gathered data from regulators about the number and nature of examination findings, where applicable.

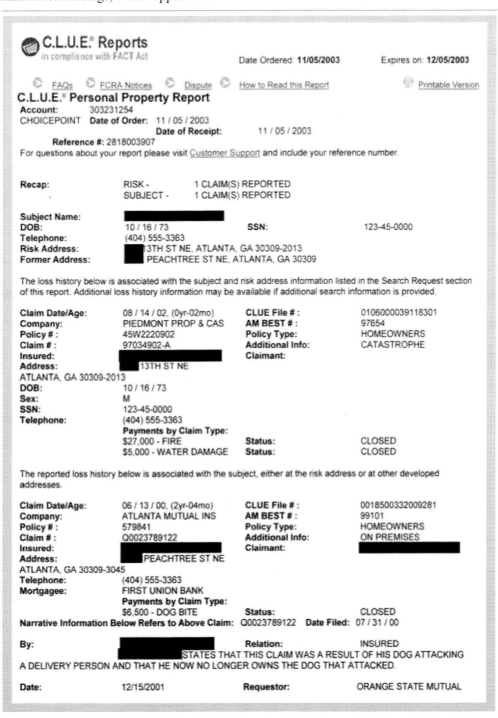

Source: ChoicePoint.

Figure 4. Sample Insurance Claims History Report.

To describe the efforts of state insurance regulators to oversee insurance companies' compliance with the Gramm-Leach-Bliley Act (GLBA), we also reviewed the DISB survey report of insurance companies' implementation of GLBA policies and procedures. DISB used the survey responses to determine findings for each company on the level of compliance with GLBA and related NAIC model rule provisions.

The DISB review defined a "finding" as an occurrence of a perceived gap between a company's privacy practices and procedures and the guidelines outlined in one of the model acts or regulations of NAIC. The findings were derived from responses to the survey questions. The companies DISB surveyed comprised major companies, including property and casualty insurance groups with 2002 gross written premiums of approximately $250 million or more; life insurance groups with 2002 gross written premiums of approximately $200 million or more; and health insurance groups with 2002 gross written premiums of approximately $500 million or more. This initial list contained 129 insurance groups. After the initial list was compiled, 26 groups were exempted from the survey examination for one of three reasons: (1) there was a prior, ongoing, or upcoming examination of the group that included (or would include) a comprehensive review of the group's privacy policy (23 groups); (2) the group engaged primarily or solely in reinsurance (2 groups); or (3) the state insurance regulator for the company's state of domicile requested that the group be exempted (1 group). The survey questionnaire included 93 questions asking for detailed documentary and testimonial evidence of companies' level of compliance with GLBA and related NAIC model rule provisions.

We conducted our review from June 2005 through May 2006 in accordance with generally accepted government auditing standards.

APPENDIX II: SAMPLE INFORMATION RESELLER REPORTS

This appendix provides examples of reports from different types of products sold by information resellers. These sample reports, which are reprinted with permission, contain fictitious data and have also been redacted to reduce possible coincidental references to actual people or places.

Sample Insurance Claims History Report

This sample insurance claims history report from ChoicePoint provides insurers with insurance claims histories on individuals applying for coverage.

Sample Deposit Account History Report

ChexSystems, a subsidiary of eFunds, offers a product that assesses risks associated with individuals applying to open new deposit accounts. The report includes information on an applicant's account history, including accounts closed for reasons such as overdrafts, returned checks, and check forgery. The report may include a numeric score representing the individual's estimated risk.

Sample Identity Verification and OFAC Screening Report

ISO, a company that provides information services to insurance companies, offers this product for screening new customers and verifying their identities. It provides a "pass" or "fail" response to indicate whether information provided by the applicant matches information maintained by the company.

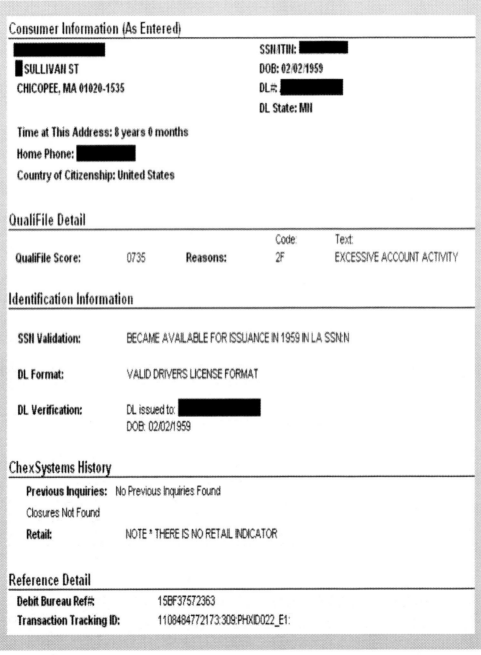

Source: eFunds.

Figure 5. Sample Deposit Account History Report.

Sample Fraud Investigation Report

Below are selected excerpts from a sample report of ChoicePoint's AutoTrack XP product, which helps users such as corporate fraud investigators and law enforcement agencies conduct investigations, locate individuals and assets, and verify physical addresses.

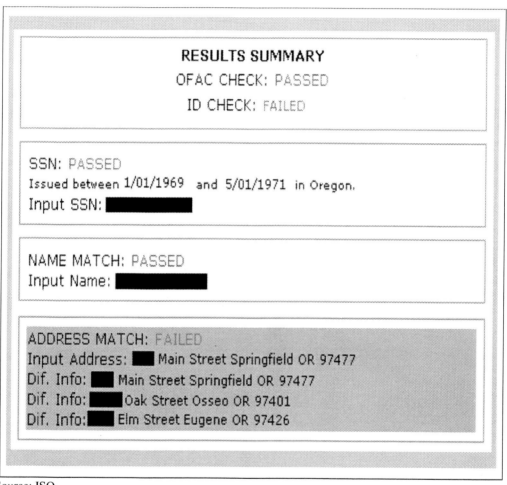

Source: ISO

Figure 6. Sample Identity Verification and OFAC Screening Report.

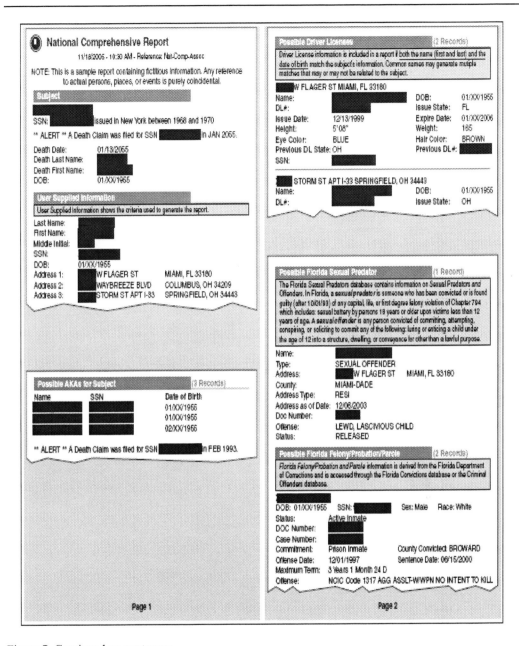

Figure 7. Continued on next page.

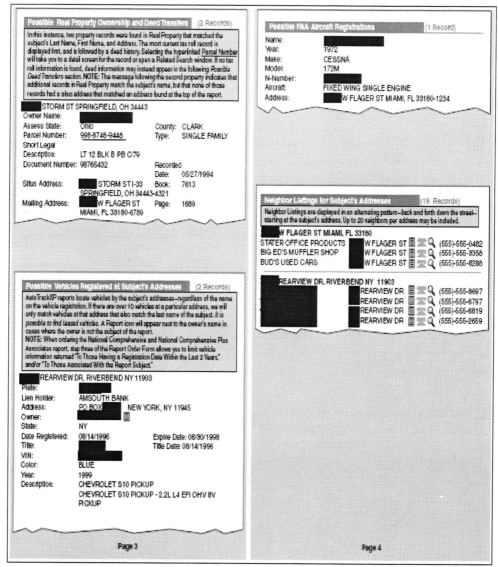

Source: ChoicePoint.

Figure 7. Sample Fraud Investigation Report.

APPENDIX III:
COMMENTS FROM THE FEDERAL TRADE COMMISSION

FEDERAL TRADE COMMISSION
WASHINGTON, D.C. 20580

THE CHAIRMAN

June 2, 2006

Ms. Yvonne Jones
Director, Financial Markets and Community Investment
Government Accountability Office
Washington, D.C. 20548

Dear Ms. Jones:

The Commission is pleased to have the opportunity to comment on the Government Accountability Office's draft report entitled: *Personal Information: Key Federal Privacy Laws Do Not Require Information Resellers to Safeguard all Sensitive Data* (GAO-06-674). ("Report") The Report describes the sources of consumer personal data, how different entities use or reuse the data, and the statutory provisions that govern the collection, use, and reuse of sensitive personal information. The Report also explains how banking regulators, the Securities and Exchange Commission, and the Federal Trade Commission ("FTC") oversee compliance with the privacy and safeguarding provisions of the Gramm-Leach-Bliley Act ("GLBA"), and describes the FTC's enforcement of the Fair Credit Reporting Act ("FCRA") with respect to information resellers. The Report concludes that "[n]o federal law explicitly requires all information resellers to safeguard all the sensitive personal information they may hold." It also finds that entities other than information resellers hold sensitive personal information.

We understand that the agencies' staffs worked cooperatively throughout the preparation of this report and that FTC staff has provided informal technical comments on the draft of the Report to the GAO staff, the vast majority of which have been incorporated.

The Report makes several legislative recommendations, two of which the Commission supports, and one on which the Commission has no opinion. First, the Report recommends that Congress consider requiring information resellers to safeguard all sensitive personal information they hold, and suggests that Congress consider the benefits and costs of expanding the class of entities explicitly required to safeguard sensitive personal information. (Report at 42) The FTC similarly has recommended that Congress consider legislative actions to increase the protection afforded sensitive personal data. In its June 2005 testimony before the Senate Committee on Commerce, Science, and Transportation on data breaches and identity theft, the Commission recommended that Congress consider extending the GLBA safeguards principles, which require

Yvonne Jones – Page 2

financial institutions to implement procedures to protect consumer financial information, to other entities that maintain sensitive information.[1]

Second, the Report recommends that Congress consider authorizing the FTC to seek civil penalties for violations of the GLBA privacy and safeguarding provisions. In its testimony to the Senate Committee cited above, the Commission noted that a civil penalty often is the most appropriate and effective remedy in cases under those provisions.[2] The Commission thus agrees with the Report's recommendation.

Finally, the Report recommends that state regulators ensure compliance with GLBA in its oversight of insurance companies. Although the Commission does not have an opinion regarding state oversight of insurance companies, the Commission agrees with GAO's conclusion that insurance companies often hold sensitive personal data.

Protecting the privacy and security of personal information collected or sold by data brokers and others is one of the Commission's highest priorities. The Commission will continue to monitor this area and will take law enforcement action when appropriate against entities that fail to protect properly sensitive consumer data.[3]

Further, the Commission encourages consumers to understand their rights under the FCRA and GLBA, and to take appropriate measures to protect their data. We have developed an array of consumer education materials for these purposes, which are available online at www.ftc.gov.

The Commission appreciates this opportunity to review and comment on GAO's Report.

By direction of the Commission.

Deborah Platt Majoras
Chairman

---

[1]   *See* Testimony of the Federal Trade Commission before the Senate Committee on Science, Commerce, and Transportation at p. 7, available at www.ftc.gov/opa/2005/06/datasectest.htm.

[2]   *Id.* at p. 9, n.18.

[3]   To date, the Commission has brought 13 legal actions against entities that allegedly failed to implement reasonable and appropriate data security for sensitive consumer data. *See* www.ftc.gov/privacy/index.html.

In: Future of the Internet: Social Networks…   ISBN: 978-1-61209-597-4
Editors: Rick D. Sullivan and Dominick P. Bartell   ©2011 Nova Science Publishers, Inc.

Chapter 4

SOCIAL SECURITY NUMBERS: INTERNET RESELLERS PROVIDE FEW FULL SSNS, BUT CONGRESS SHOULD CONSIDER ENACTING STANDARDS FOR TRUNCATING SSNS*

United States Government Accountability Office

WHAT GAO FOUND

We found 154 Internet information resellers with SSN-related services. Most of these resellers offered a range of personal information, such as dates of birth, drivers' license information, and telephone records. Many offered this information in packages, such as background checks and criminal checks. Most resellers also frequently identified individuals, businesses, attorneys, and financial institutions as their typical clients, and public or nonpublic sources, or both as their sources of information.

In attempting to purchase SSNs from 21 of the 53 resellers advertising the sale of such information, we received 1 full SSN, 4 truncated SSNs displaying only the first five digits, and no SSNs from the remaining 16. In one case, we also received additional unrequested personal information including truncated SSNs of the search subject's neighbors. We also found that some other entities truncate SSNs by displaying the last four digits. According to experts we spoke to, there are few federal laws and no specific industry standards on whether to display the first five or last four digits of the SSN, and SSA officials told us the agency does not have the authority to regulate how other public or private entities use SSNs, including how they are truncated.

We could not determine if federal privacy laws were applicable to the Internet resellers because such laws depend on the type of entity and the source of information, and most of the resellers' Web sites did not include this information. However, these laws could apply to

---

* This is an edited, reformatted and augmented edition of a United States Government Accountability Office publication, Report GAO-06-495, dated May 2006.

resellers; 4 of the resellers we examined had Web sites identifying the type of entity they were. About one-half of the resellers cited adherence to one or more federal privacy laws and a few referenced state laws.

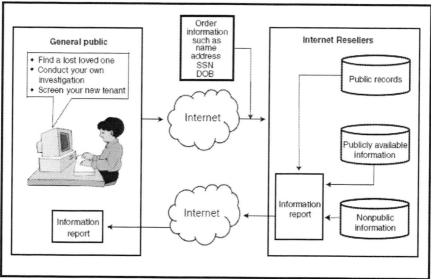

Source: GAO analysis.

How the General Public Can Purchase Information from Internet Resellers.

WHY GAO DID THIS STUDY

GAO previously reported on how large information resellers like consumer reporting agencies obtain and use Social Security numbers (SSNs). Less is known about information resellers that offer services to the general public over the Internet. Because these resellers provide access to personal information, SSNs could be obtained over the Internet. GAO was asked to examine (1) the types of readily identifiable Internet resellers that have SSN-related services and characteristics of their businesses, (2) the extent to which these resellers sell SSNs, and (3) the applicability of federal privacy laws to Internet resellers.

WHAT GAO RECOMMENDS

Since there is no consistently practiced method for truncating SSNs and no federal agency has the authority to regulate how SSNs could be truncated, Congress may wish to consider enacting standards for truncating SSNs or delegating authority to the Social Security Administration (SSA) or some other governmental entity to issue standards for truncating SSNs. In commenting on a draft of this chapter, SSA agreed that standardizing the truncation of SSNs would be beneficial and supported our recommendation for congressional action.

ABBREVIATIONS

DCI	data collection instrument
DPPA	Driver's Privacy Protection Act
FACTA	Fair and Accurate Credit Transactions Act
FCRA	Fair Credit Reporting Act
FTC	Federal Trade Commission
GLBA	Gramm-Leach-Bliley Act
MSN	Microsoft Network
SSA	Social Security Administration
SSN	Social Security number

May 17, 2006

The Honorable Jim McCrery Chairman
Subcommittee on Social Security Committee on Ways and Means House of Representatives

The Honorable E. Clay Shaw, Jr. House of Representatives

The Social Security number (SSN) is a key piece of personal information and has come to be used for numerous non–Social Security purposes. In recent years, both public and private sector entities have increasingly used the SSN as a personal identifier and ask individuals to supply their SSNs. Consequently an individual's SSN can be found on a number of public documents such as land ownership records, birth certificates, and marriage licenses, and is advertised for sale. Private-sector entities known as information resellers are specializing in amassing personal information, including SSNs, from various public and private sources and providing information about someone for specific purposes for a fee.

More prominent or large information resellers limit their services to businesses and government entities that establish accounts with them and have a legitimate purpose for obtaining personal information on an individual. However, less is known about other information resellers, particularly those that are Internet-based and offer their services to the public at large for a fee. Such Internet information resellers (Internet resellers) make public and nonpublic information accessible to the public, raising concerns about how easy it would be for someone to obtain another person's SSN over the Internet. At your request, we (1) describe the types of readily identifiable Internet resellers that have SSN-related services and characteristics of their business, (2) determine the extent to which these Internet resellers sell SSNs, and (3) determine the applicability of federal privacy laws to Internet resellers.

To identify Internet resellers and their characteristics, we developed an initial list of over 1,000 potential Internet resellers by searching the Internet with popular Web-based search engines, such as Google, and using keywords and phrases that members of the general public would use if they were trying to find Web sites that would allow them to obtain someone else's SSN and other personal information. We narrowed the list of Internet resellers to 154 distinct Web sites that had services that either required the customer to provide the reseller with an SSN or sold an SSN. We then used a data collection instrument (DCI) to capture information posted on resellers' Web sites about their characteristics, such as the types of information available for sale, the types of clients resellers market to, and the sources of information they stated they used. To determine the extent to which the Internet resellers sell SSNs, we analyzed the data obtained from the DCI about Internet resellers with SSN-related services and attempted to purchase the SSNs of consenting GAO staff members from a nonprobability sample of 21 resellers on the list. [1] The criteria we used to select the resellers for our attempted purchases included (1) Web sites that advertise the sale of an SSN without the customer's having to provide the SSN of the subject of our inquiry, (2) Web sites that advertise the sale of an SSN to the general public, and (3) the Web sites where the transaction could be made online through use of a credit card. We also interviewed staff from the Federal Trade Commission (FTC), officials from the Social Security Administration (SSA), industry representatives, and privacy experts to get their views about the use of SSN truncation. To determine the applicability of federal privacy laws to the Internet resellers, we reviewed federal privacy laws and examined pertinent information on the resellers' Web sites, including their references to privacy laws. Appendix I explains the scope and methodology of our work in greater detail. We conducted our work between April 2005 and May 2006 in accordance with generally accepted government auditing standards.

RESULTS IN BRIEF

Although numerous Internet resellers exist, resellers' Web sites we reviewed generally had similar characteristics. Most advertised a selection of personal information ranging from previous and current addresses and dates of birth to drivers' license information, telephone records, and credit reports. In addition, many of them offered to sell personal information in various packages, such as criminal checks and background checks. Web sites most frequently identified individuals, businesses, attorneys, and financial institutions as their typical clients and public or nonpublic sources, or both as their sources of information.

We generally failed in our attempts to purchase full SSNs, although we did receive other personal information. Of the 53 Web sites that offered to sell a person's SSN, we tried to purchase SSNs of consenting GAO employees from 21 of these resellers and received one complete SSN for the person whose number we requested; four truncated SSNs, where only the first five digits were disclosed (123-45-XXXX); and no SSN from the remaining 16. In our discussions with privacy experts, private sector representatives, and federal officials, we found that entities in other industries, such as credit reporting, sometimes truncate the SSN by masking the first five digits of the SSN but displaying the last four (XXX-XX-1234). These experts added there are few federal laws, and no specific industry standards, about which digits of an SSN are displayed in a truncated format. According to SSA officials, SSA does not have the authority to regulate how other public and private entities use SSNs, including how they are truncated. Furthermore, when we were successful in purchasing truncated SSNs as part of a background check, we also received personal information such as an individual's address, date of birth, and list of neighbors. In one case, we received unrequested information including the truncated SSNs of the person's current and past neighbors.

We could not determine if federal privacy laws were applicable to the Internet resellers because such laws depend on the type of entity involved and the source of information, and most of the resellers' Web sites did not include this information. Certain federal privacy laws—such as the Fair Credit Reporting Act (FCRA), the Gramm-Leach-Bliley Act (GLBA), and the Driver's Privacy Protection Act (DPPA)—restrict the disclosure of personal information based on the type of entity or the specific source of the information. We found that most of the Internet resellers' Web sites we reviewed had insufficient information on their Web sites for us to determine the type of entity the reseller was or the source of the reseller's information. However, federal privacy laws could apply to these resellers.

In four cases, we found that the resellers stated on their Web sites the type of entity they were—consumer reporting agencies and credit bureau— which are subject to FCRA or GLBA. We also found that about 79, or one- half, of the resellers referenced one or more federal privacy laws on their Web sites, indicating some awareness of these laws, while others referenced certain state laws, such as those of California, Florida, and Michigan.

Because different entities truncate SSNs in different ways and no federal agency has the authority to regulate how SSNs should be truncated, Congress may wish to consider enacting standards for truncating SSNs or delegating that authority to SSA or some other governmental agency.

In commenting on a draft of this chapter, SSA agreed that standardizing the truncation of SSNs would be beneficial and supported our recommendation to Congress.

BACKGROUND

The SSN was created in 1936 as a means of tracking workers' earnings and eligibility for Social Security benefits. SSNs are issued to most U.S. citizens, and to some noncitizens lawfully admitted to the United States. Through a process known as enumeration, a unique nine-digit number is created. The number is divided into three parts— first three digits represent the geographic area where the SSN was assigned; the middle two are the group number, which is assigned in a specified order for each area number; and the last four are serial numbers ranging from 0001 to 9999. Because of the number's uniqueness and broad applicability, SSNs have become the identifier of choice for government agencies and private businesses, and are used for a myriad of non–Social Security purposes.

Information resellers, sometimes referred to as information brokers, are businesses that specialize in amassing personal information from multiple sources and offering informational services. These entities may provide their services to a variety of prospective buyers, either to specific business clients or to the general public through the Internet. More prominent or large information resellers such as consumer reporting agencies and entities like LexisNexis provide information to their customers for various purposes, such as building consumer credit reports, verifying an individual's identity, differentiating records, marketing their products, and preventing financial fraud.

These large information resellers limit their services to businesses and government entities that establish accounts with them and have a legitimate purpose for obtaining an individual's personal information. For example, law firms and collection agencies may request information on an individual's bank accounts and real estate holdings for use in civil proceedings, such as a divorce.

Information resellers that offer their services through the Internet (Internet resellers) will generally advertise their services to the general public for a fee.

Resellers, whether well-known or Internet-based, collect information from three sources: public records, publicly available information, and nonpublic information.

- Public records are available to anyone and obtainable from governmental entities. Exactly what constitutes public records depends on state and federal laws, but generally includes birth and death records, property records, tax lien records, voter registrations, and court records (including criminal records, bankruptcy filings, civil case files, and legal judgments).
- Publicly available information is information not found in public records but nevertheless available to the public through other sources. These sources include telephone directories, business directories, print publications such as classified ads or magazines, and other sources accessible by the general public.
- Nonpublic information is derived from proprietary or private sources, such as credit header data [2] and application information provided by individuals—for example, information on a credit card application—directly to private businesses.

Table 1. Aspects of Selected Federal Laws Affecting Public and Private Sector Disclosure of Personal Information

Federal laws	Restrictions on disclosure	Entities affected
Gramm-Leach-Bliley Act (GLBA)	Creates a new definition of nonpublic personal information that includes SSNs and gives consumers the right to limit some, but not all, sharing of their nonpublic personal information. Financial institutions can disclose consumers' nonpublic information without offering them an opt-out right under certain circumstances permissible under the law, such as to protect the confidentiality or security of the consumer's record and to prevent actual or potential fraud.	Financial institutions such as credit bureausand entities that receive data from financial institutions
Fair Credit Reporting Act (FCRA)	Limits access to consumer reports, which generally include SSNs, to those who have a permissible purpose under the law, such as state or local officials involved in the enforcement of child support cases or determining eligibility for employment.	Consumer reporting agencies and users of consumer reports
Fair and Accurate Credit Transactions Act (FACTA)	Amends FCRA to allow, among other things, consumers who request a copy of their credit report to also request that the first five digits of their SSN (or similar identification number) not be displayed; requires consumer reporting agencies and any business that uses consumer reports to adopt procedures for proper disposal of such reports.	Consumer reporting agencies and users of consumer reports
Driver's Privacy Protection Act (DPPA)	Prohibits disclosing personal information from a motor vehicle record, including SSNs, except for purposes permissible under the law.	State departments of motor vehicles, department of motor vehicle employees or contractors, and recipients of personal information from motor vehicle records

Source: GAO analysis.

Information resellers provide information to their customers for various purposes, such as building consumer credit reports, verifying an individual's identity, differentiating records, marketing their products, and preventing financial fraud. The aggregation of the general public's personal information, such as SSNs, in large corporate databases and the increased availability of information via the Internet may provide unscrupulous individuals a means to acquire SSNs and use them for illegal purposes.

Because of the myriad uses of the SSN, Congress has previously asked GAO to review various aspects of SSN-use in both the public and private sectors [3].

In our previous work, our reports have looked at how private businesses and government agencies obtain and use SSNs [4]. In addition, we have reported that the perceived widespread sharing of personal information and instances of identity theft have heightened public concern about the use of Social Security numbers [5].

We have also noted that the SSN is used, in part, as a verification tool for services such as child support collection, law enforcement enhancement, and issuing credit to individuals [6]. Although these uses of SSNs are beneficial to the public, SSNs are also key elements in creating false identities. We testified before the Subcommittee on Social Security, House Committee on Ways and Means, about SSA's enumeration and verification processes and also reported that the aggregation of personal information, such as SSNs, in large corporate databases, as well as the public display of SSNs in various public records, may provide criminals the opportunity to commit identity crimes [7].

We have also previously reported that certain federal and state laws help information resellers limit the disclosure of personal information including SSNs to their prescreened clients. [8] Specifically, we described how certain federal laws place restrictions on how some Internet resellers' obtain, use, and disclose consumer information. The limitations these laws afford are shown in table 1.

INTERNET RESELLERS' WEB SITES SHARED SIMILAR CHARACTERISTICS

The Web sites of the 154 Internet resellers we reviewed had similar characteristics. Most resellers offered a variety of information that could be purchased, from telephone records to credit reports. In addition, Internet resellers also offered to sell information in various ways, from packaged information, such as various information that would be collected through a background check or a search of a person's criminal records to single types of information, such as a credit score. These resellers usually listed the types of clients that they market their services to and broadly identified their sources of information.

Internet Resellers Offered to Sell a Variety of Information in Various Ways

We found that Internet resellers offered to sell a variety of information to anyone willing to pay a fee. On average, resellers offered about 8 types of services and two offered 20 types of informational services. As shown in figure 1, the majority of resellers offered to sell anywhere from 1 to 10 informational services.

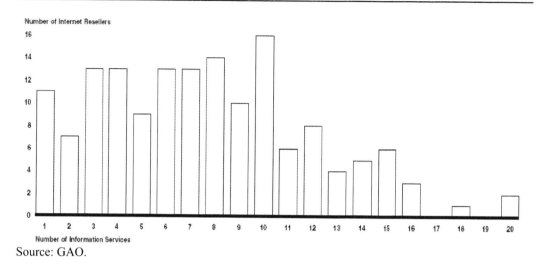

Source: GAO.

Figure 1. Number of Services Provided by the 154 Internet Resellers.

The Internet resellers offering the fewest services tended to specialize in services provided to the public. For example, most of the resellers offering only one service were resellers that specialized in helping locate an individual. Others offered services related to employment or background checks.

Internet resellers also offered different ways for buyers to purchase their information. For example, some offered memberships that allowed online access to the reseller's information, with the member performing the search. Another reseller offered to sell a software package that would allow a buyer to purchase access to the Internet reseller's information through the purchased software and allowed many different types of information searches. The majority of resellers would require selected information about the buyer and then would perform the data search and provide an information report to the buyer.

We identified over 50 types of information offered for purchase by these resellers, which we categorized into six major categories including personal, legal, financial, employment, driver or vehicle, and telephone. Table 2 gives examples of the types of information found in these categories.

Table 2. Categories and Examples of Information Provided by Internet Resellers

Information categories	Types of Information in these categories
Personal	Name, SSN, aliases, current and previous addresses, telephone number, and date of birth or age
Legal	Federal, state, and county criminal records checks
Financial	Credit reports, credit cards, bank accounts, and bankruptcy records search
Employment	Employment history and salary or income verification
Driver or vehicle	Driver's license number and driver's history report
Telephone	Telephone and cell phone records and name and address ofan individual based on his telephone or cell phone number

Source: GAO analysis.

All the resellers offered to sell information from at least one of the six categories. However, not all resellers offered to sell driver or vehicle information, or telephone information. For example, only 85 of the 154 resellers we reviewed offered to sell some type of driver's information, while 56 resellers offered to sell telephone information.

We found that Internet resellers either sold their information as a part of a package or sold single pieces of information. For example, resellers sold packaged information such as background checks, criminal checks, or employment checks/tenant screenings. Of the packaged information, we found that background checks provided the most extensive information. A background check may include personal, legal, and financial information, such as name, SSN, address, neighbors, relatives, and associates information. Such checks may include national, state, or county criminal records searches and bankruptcy and lien information. [9] Other packages, such as criminal records packages, may include national, state, and county criminal records searches, sex offender searches, and civil litigation. Employment checks/tenant screenings may include current and past employment, SSN verifications, and national, state, and county criminal records searches.

Internet Resellers Usually Identified Their Clients

Over 80 percent of Internet resellers identified the clients to whom they marketed their information. Internet resellers identified their clients in several ways. About 60 percent of the time, resellers used the information sections of their Web sites to identify their clients. Web pages such as "Frequently Asked Questions," "Help," or "About Us" were frequently used to identify their clients. For example, the "About Us" Web page generally provided a brief description about the Internet reseller's business and would often describe the clients it marketed to. Other ways in which resellers marketed to their clients were through testimonials or in a separate section on their Web page.

Internet resellers marketed their services to a variety of clients. As shown in table 3, individuals, businesses, and attorneys were the most frequently identified clients. Some of the businesses resellers identified were Fortune 500 companies and retailers.

Table 3. Types of Clients to Which Internet Resellers Market Their Services

Types of clients	Internet resellers that marketed to these clients
Individuals	84
Businesses	72
Attorneys	42
Financial institutions	29
Insurance agents or agencies	26
Private investigators	23
Government or law enforcement agencies	21
Collection agencies	12
Landlords	11
Health services	8
Other	16

Source: GAO analysis.

For the financial institution clients, resellers mostly identified banks. In addition, most of the Internet resellers' clients were from the private sector, although some had government and law enforcement agency clients. Finally, we found that most of the resellers had multiple types of clients. About 30 percent of the resellers identified only one type of client.

Three-Quarters of Internet Resellers Identified Their Sources of Information

About 75 percent, or 115, Internet resellers identified the source of their information on their Web sites. Most of these resellers obtained their information from public or nonpublic sources or a combination of both sources. For example, a few resellers offered to conduct a background investigation on an individual, which included compiling information on the individual from court records and using a credit bureau to obtain consumer credit data. Some used only public records as their only source of information. The most frequently identified public records were court records, department of motor vehicle records, real property records, legal judgments, and bankruptcy records. We found about one-third of the Internet resellers used only one source of information. More often, they used a combination of the three sources. Figure 2 shows the various combinations of sources of information.

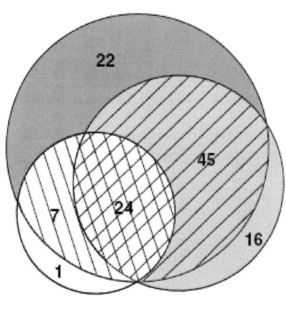

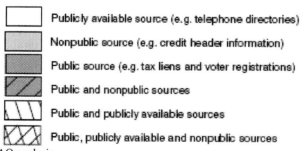

Source: GAO analysis.

Figure 2. Combinations of the Sources of Information Used by Internet Resellers.

MOST ATTEMPTS TO PURCHASE SSNS FAILED

Most of our attempts to purchase SSNs from a select group of resellers failed. Of the 154 Internet resellers' Web sites we reviewed, 53, almost 35 percent, offered to sell SSNs. We attempted to purchase SSNs from 21 resellers that were chosen because they required minimal information about prospective buyers or about the person whose SSN we wanted to obtain. Of the 21 resellers from which we tried to purchase SSNs, only 5 provided some form of an SSN. As shown in table 4, the reasons for being unable to obtain SSNs from 16 of the 21 resellers varied.

Nine resellers, a majority of the resellers that did not sell SSNs to us, did not explain why but simply did not provide the information we sought. Four of the remaining resellers attempted to contact us to request legal documentation to support a permissible purpose for obtaining the information. However, since we attempted to purchase SSNs as a member of the general public, we could not provide the requested information. One of these resellers sent us an e-mail asking us to fax a signed letter stating our reason for obtaining a person's SSN and a copy of our driver's license to verify our identity, which we could not provide. We contacted the other three to find out why prospective buyers were required to have a permissible purpose. One reseller told us that the company is audited every year by the government and that a legal document request was part of its security screening of its customers. The other two stated that some form of legal documentation, such as a certified copy of a court order, was required in order for their companies to release the information.

In addition to receiving one full and four truncated SSNs, we also received other information related to our purchases. Given that we only received SSNs as a part of packaged information, we were not surprised that we received additional information about the person whose SSN we were trying to obtain. For example, the two Internet resellers that provided some form of SSN in a background check report also provided the following information:

- the person's current and previous addresses,
- date of birth,
- a list of other names associated with the person,
- a list of their neighbors,
- tax liens and judgments against the person, and
- properties owned by the person [10].

However, in one case we received unexpected and unrequested information. In this case, we did not receive the SSN of the person whose number we requested, but instead received the truncated SSNs of the person's past and present neighbors, information we did not request.

Five of the 21 resellers from whom we attempted to purchase SSNs did provide us with some form of an SSN. We received one full nine-digit SSN and four truncated SSNs. All five resellers that supplied an SSN provided the SSN as a part of a package of information. As shown in table 5, the full SSN was obtained as a part of a background check, and the four truncated SSNs were provided as a part of a "people locator" package, a background package, and an employment trace. We attempted to order SSNs from five resellers that offered to sell the SSN alone, and we were unable to obtain an SSN from those resellers.

Table 4. Reasons Internet Resellers Did Not Provide SSNs

Reason	Internet reseller
Required additional legal documentation of permissible purpose for obtaining the information	4
Refused because of state privacy laws	1
Required forms of payment other than a credit card	1
No record found on subject	1
Reason unknown	9
Total	16

Source: GAO analysis.

Table 5. Results of Attempted SSN Purchases

SSN services	Orders placed[a]	Received full SSN	Received truncated SSN
SSN alone (e.g., Locate an SSN, search for Social Security numbers, and SSN search)	5	0	0
Background check or investigation	6	1	1
People locator or search	5	0	2
Employment trace	1	0	1
Other information packages	4	0	0
Total	21	1	4

Source: GAO analysis.

[a]Does not include three attempted orders where we received an error message after submitting our information that terminated our transaction.

We also found a wide range of the costs for information services when we tried to purchase SSNs. The packages of information we attempted to purchase ranged from about $4 to $200 compared to the costs to purchase individual SSNs that ranged from about $15 to $150. The range of costs from the five resellers that provided some form of the SSN was about $20 to $200. The Internet reseller that provided the full SSN did so for $95.

Of the four resellers that gave us truncated SSNs, three of these disclosed on their Web sites that they would provide full SSNs, but only under certain circumstances. For example, one reseller said that, by law, it cannot provide a person's SSN to any third party. Another required the customer to have a legitimate reason for requesting the information under laws such as GLBA. This reseller said it may not provide the full SSN if the customer did not meet those requirements. None explained why they only provided the first five digits.

All resellers that provided truncated SSNs showed the first five digits and masked the last four digits. We interviewed industry representatives and privacy experts to determine if this way of truncating the SSN was the standard practice among private sector entities. Industry representatives and privacy experts told us that entities in other industries may truncate the SSN differently from the truncated SSNs we bought from Internet resellers. For example, consumer data industry representatives said that members of their association decide for themselves how and when to truncate SSNs. One consumer reporting agency we spoke to told us that it truncates the SSN by masking the first five digits on reports it provides directly to consumers, by displaying only the last four digits. Some privacy experts said that certain entities that use SSNs as identifiers on lists, such as universities, also truncate the number by masking the first five digits. In addition, SSA also masks the first five digits of the SSN on the Social Security Statements mailed to individuals over the age of 25 who have an SSN and have wages or earnings from self–employment.

On the basis of our discussions with government officials and industry representatives, we could not identify any industry standards or guidelines for truncating SSNs. None of the officials we spoke to knew for certain why either method—masking the first five digits or the last four digits— was used or how such methods came into use. In addition, when we asked officials which way of truncating the SSN better protects it from misuse, there was no consensus among them, and no one knew of any research regarding this issue. Some officials said that although truncation could provide some protection for SSNs, it is unlikely to be foolproof. There are also few, if any, federal laws that require or regulate truncating the SSN. Currently, FCRA has a specific provision relating to truncating SSNs. Under this law consumers can request that their SSN be truncated to display only the last four digits on any consumer report they request about themselves. The Judicial Conference of the United States issued rules, effective in December 2003, requiring that SSNs be truncated to mask the first five digits in newly filed electronically available bankruptcy court documents.

Federal agency officials whom we spoke to said that Congress or SSA should decide how SSNs should be truncated. The Social Security Act of 1935 authorized SSA to establish a record-keeping system to help manage the Social Security program and resulted in the creation of the SSN. Through a process known as enumeration, unique numbers are created for every person as a work and retirement benefit record for the Social Security program. According to SSA officials, the law does not address the use of the number by private and public sector entities. SSA officials said that SSA regulates only the agency's use of SSNs and does not have legal authority over SSNs used by others.

APPLICABILITY OF FEDERAL PRIVACY LAWS TO INTERNET RESELLERS CANNOT BE DETERMINED

Federal privacy laws that restrict the disclosure of personal information could be applicable to Internet resellers, but there was insufficient evidence on the resellers' Web sites we reviewed to determine if they met specific statutory definitions. Federal privacy laws such as the FCRA, GLBA, and DPPA apply primarily to entities that meet specific statutory definitions. For example, FCRA applies primarily to a consumer reporting agency, which is defined as any person which, for monetary fees, dues, or on a cooperative nonprofit basis, regularly engages in whole or in part in the practice of assembling or evaluating consumer credit information or other information on consumers for the purpose of furnishing consumer reports to third parties, and which uses any means or facility of interstate commerce for the purpose of preparing or furnishing "consumer reports" [11]. In addition, these laws allow for disclosure of personal information for certain permissible purposes, and those who request or receive information from an entity meeting those statutory definitions may also have obligations under these laws. For example, FCRA generally prohibits "consumer reporting agencies" from furnishing "consumer reports" to third party users unless it is for a permissible purpose; before providing "consumer report" information to prospective users, however, the prospective user must certify the purposes for which the information is sought and that it will be used for no other purpose. [12] GLBA and DPPA also contain prohibitions against re-disclosure of personal information covered by those laws. [13]

FCRA, GLBA, and DPPA could apply to Internet resellers that identify themselves as one of the statutorily defined entities covered under the laws—which are consumer reporting agencies for FCRA, financial institutions for GLBA, and state motor vehicle departments for DPPA—or that received information from such entities. We found four resellers that identified themselves as one of the statutorily defined entities. Three stated on their Web sites that they were consumer reporting agencies and the other stated it was a credit bureau. However, we did not find similar information on the remaining 150 resellers' Web sites to determine what type of entity they were. In addition, we found that some resellers identified the source of their information generally, but did not link information sources to particular pieces of information. For example, about 7 percent of the resellers identified "Department of Motor Vehicle records" as the source of some of their information and offered to search for personal information based on a driver's license number, license plate number, or vehicle identification number. However, most did not specify which personal information came from the "Department of Motor Vehicle records" or any state motor vehicle departments. Therefore, we could not determine if FCRA, GLBA, and DPPA were applicable to the majority of resellers we reviewed.

Our review of the resellers' Web sites found 79 of them, about 50 percent, referenced one or more federal privacy laws. As shown in figure 3, the most frequently mentioned laws were FCRA, GLBA, and DPPA.

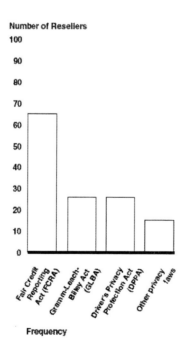

Source: GAO analysis.

Figure 3. Frequency of Federal Privacy Laws Cited by Internet Resellers.

We also found 5 out of the 154 Internet resellers referenced state laws on their Web sites. Two stated adherence to the California Investigative Consumer Reporting Act, which allows a consumer to review any files concerning that consumer maintained by an "investigative reporting agency." One cited two California consumer laws. One law allows California consumers to remove their names from credit bureau mailing lists used for unsolicited pre-approved credit offers for a minimum of 2 years. It also provides identity theft victims and other consumers with increased rights regarding consumer credit reports, including requiring the deletion of inquiries resulting from identity theft. The other California law prohibits consumer credit reporting agencies that furnish reports for employment purposes from reporting information on the age, marital status, race, color, or creed of any consumer and requires the user of the report to provide written notice to the consumer. The law also requires that the consumer be provided a free copy of the report upon request. Another reseller cited a Florida statute that governs divulging investigative information, and yet another reseller stated adherence to the Michigan Private Detective License Act. Both state laws regulate the activities of private investigators.

CONCLUSIONS

Although personal information is widely available on the Internet to anyone willing to pay a fee, SSNs appear to be difficult to obtain from the Internet resellers we contacted. Few of the Internet resellers' Web sites we reviewed offered to sell an individual's SSN outright, and even those that did make such an offer did not follow through. Thus, the perception that anyone willing to pay a fee can easily obtain someone's SSN does not appear to be valid. Our experiences indicate that it is more likely that a buyer would not be able to purchase an SSN or would receive a truncated version of an SSN from Internet resellers.

However, our work does suggest that someone seeking an SSN may be able to obtain a truncated SSN, and depending on the entity, the SSN may be truncated in various ways. Standardizing the truncation of the SSN could provide some protection from SSNs being misused. Under a standardized approach, the same digits of the SSN would be the only information transmitted, no matter the source from which the SSN is obtained. Given SSA's role in assigning SSNs, SSA is in the best position to determine whether and if truncation should be standardized, but because the agency does not have specific authority to regulate truncation, SSN truncation will continue to vary.

MATTER FOR CONGRESSIONAL CONSIDERATION

Since there is no consistently practiced method for truncating SSNs, and no federal agency has the authority to regulate how SSNs should be truncated, Congress may wish to consider enacting standards for truncating SSNs or delegating authority to SSA or some other governmental entity to issue standards for truncating SSNs.

AGENCY COMMENTS AND OUR EVALUATION

We provided a draft of this chapter to the Social Security Administration for comment and received a written response from the administration (included as app. II). SSA agreed that standardizing the truncation of SSNs would be beneficial and supported our recommendation for congressional action. In addition, SSA stated that while it does not have the legal authority to compel organizations to truncate SSNs or to specify how such truncating should be done, it would be willing to publish information on best practices for truncating SSNs on SSA's Web site. We also provided a draft of this chapter to the Federal Trade Commission for technical review and received comments that were incorporated as appropriate.

REFERENCES

[1]   We selected these Web sites using a nonprobability sample—a sample in which some items in the population have no chance, or an unknown chance, of being selected. Results from nonprobability samples cannot be used to make inferences about a population. Thus, the information we obtained cannot be generalized to the other Web sites we studied.

[2]   Credit header data consist of the nonfinancial identifying information located at the top of a credit report, such as name, current and prior addresses, telephone number, Social Security number, and date of birth.

[3]   See GAO, Social Security Numbers: Government Benefits from SSN Use but Could Provide Better Safeguards, GAO-02-352 (Washington, D.C.: May 31, 2002), and Identity Theft: Prevalence and Cost Appear to Be Growing, GAO-02-363 (Washington, D.C.: Mar. 1, 2002).

[4]   GAO, Social Security: Government and Commercial Use of the Social Security Number Is Widespread, GAO/HEHS-99-28 (Washington, D.C.: Feb. 16, 1999).

[5]   GAO, Social Security: Government and Other Uses of the Social Security Number Are Widespread, GAO/T-HEHS-00-120 (Washington, D.C.: May 18, 2000).

[6]   GAO/HEHS-99-28.

[7]   GAO, Social Security Numbers: Ensuring the Integrity of the SSN, GAO-03-941T (Washington, D.C.: July 10, 2003).

[8]   GAO, Social Security Numbers: Private Sector Entities Routinely Obtain and Use SSNs, and Laws Limit the Disclosure of This Information, GAO-04-11 (Washington, D.C.: January 22, 2004).

[9]   A lien is a charge upon real or personal property for the satisfaction of some debt or duty.

[10]   The list of personal information represents some of the information the two resellers provided in background check reports.

[11]   15 U.S.C. § 1681a(f). FCRA defines a "consumer report" as any written, oral, or other communication of "any information by a consumer reporting agency bearing on a consumer's credit worthiness, credit standing, credit capacity, character, general reputation, personal characteristics, or mode of living which is used or expected to be used or collected in whole or in part for the purpose of serving as a factor in

establishing the consumer's eligibility for: (1) credit or insurance to be used primarily for personal, family, or household purposes; (2) employment purposes; or (3) any other purpose authorized under section 1681b of this title." 15 U.S.C. § 1681a(d).

[12] 15 U.S.C. § 1681e.

[13] 15 U.S.C. § 6802(c); 18 U.S.C. § 2721(c).

[14] live site is a Web site that is currently in operation and offers online services. The Web sites were live when GAO analyst reviewed the uniform resource locator (URL) for the survey. Those Web sites considered not live displayed an error message noting that the Web site was no longer in operation.

[15] 2redirected Web site is a site that acts as a portal to other Web sites. Several reseller Web sites have links to other individual reseller sites. For this survey, we reviewed the individual reseller sites and not the portal sites.

APPENDIX I: SCOPE AND METHODOLOGY

To describe readily identifiable Internet resellers, we created a list of Internet reseller Web sites. To create a list of readily identifiable Internet reseller Web sites, we used Internet search techniques and keyword search terms that we thought the members of general public would use if they were trying to obtain someone else's Social Security Number (SSN). We conducted our searches using three major Internet search engines— Google, Microsoft Network (MSN), and Yahoo. Within each of these search engines we conducted our searches using keywords such as, "find social security number," "find ssn," "purchase social security number," and "public records search." We chose these keywords based on the advice of privacy experts and the team's judgment on terms that would yield Web sites that sell personal information including the SSN.

Our searches resulted in 1,036 Web sites that we then reviewed to determine whether they were live sites, [14] redirected sites, [15] or duplicate sites that were operated by the same reseller. Nineteen percent of the 1,036 Web sites took us to another Internet reseller Web site that was included in our list. Most of these redirected sites took us to two Internet resellers that offered online membership—allowing access to their databases and affiliate programs, which allowed others to link their Web sites to the resellers' Web sites. More than one-half of the 1,036 Web sites were inactive at the time a GAO analyst attempted to access the site.

In addition, we found a few Web sites were operated by the same reseller and were similar in appearance. As a result, we ended up with a list of 226 sites that we included in our review. We recognize that had we used different search engines, different keywords, and a different point in time we may have identified a different list of sites.

To describe the types of readily identifiable Internet resellers that have SSN-related services and characteristics of their businesses, we developed a Web-based data collection instrument (DCI) for GAO analysts to document selected information contained on the Internet resellers' Web sites. We used the DCI to record information from the Web pages that contained items that addressed the types of SSN-related services and information that the resellers sold, the sources of the information, and the types of clients to whom the site marketed.

To ensure that the entry of the DCI data conformed to GAO's data quality standards, each DCI was reviewed by one of the other GAO analysts. Tabulations of the DCI items were automatically generated from the Web-based DCI software. Supplemental analyses were conducted using a statistical software package. For these analyses, the computer programs were checked by a second, independent analyst. Our analyses found 154 Internet resellers with SSN-related services.

To determine the extent to which Internet resellers sell Social Security numbers, we analyzed data collected from the review of Internet resellers just described, attempted to purchase SSNs from a nonprobability sample of Internet resellers, and collected data about the transactions. We used information collected from the DCI to derive a nonprobability sample of Internet resellers to purchase SSNs.

The criteria we used to select the resellers for our attempted purchases included the following (1) the Web site advertised the sale of an SSN without the customer's having to provide the SSN of the subject of our inquiry, (2) the Web site advertised the sale of an SSN to the general public, and (3) the transaction could be made online through the Internet reseller's Web site using a credit card. We collected information about the purchases including cost, the information that was required about the search subject and the purchaser (including the permissible purpose), whether the site contacted us to verify our information or our permissible purpose, and whether the SSN was provided and, if it was, whether the full or a truncated SSN was provided.

In addition, we interviewed staff from the Federal Trade Commission, officials from the Social Security Administration, one of the three national consumer reporting agencies, the Consumer Data Industry Association (an international trade association that represents consumer information companies), and five privacy experts to obtain their views about the use of SSN truncation as a means for safeguarding the number. We also reviewed prior GAO work and performed literature and Internet searches about SSN truncation.

To determine the applicability of federal privacy laws to Internet resellers, we reviewed federal laws and the resellers' Web sites for information about the resellers' type of entity and sources of information. However, in most instances these resellers did not have sufficient information on their Web sites to determine if they were in compliance with these laws. Specifically, we were unable to determine whether most of these resellers met the definitions specified by these laws such as "financial institution," "consumer reporting agency," or an "officer, employee, or contractor" of a "State Motor Vehicle Department." We also were unable to determine the resellers' specific sources for particular pieces of information. Although Internet resellers generally did not provide information about the entity and sources of information, they generally cited, and we recorded, whether they stated adherence to any federal privacy laws.

APPENDIX II: COMMENTS FROM THE SOCIAL SECURITY ADMINISTRATION

SOCIAL SECURITY
The Commissioner

May 05, 2006

Ms. Barbara D. Bovbjerg
Director, Education, Workforce,
  and Income Security Issues
Room 5968
U.S. Government Accountability Office
Washington, D.C. 20548

Dear Ms. Bovbjerg:

Thank you for the opportunity to review and comment on the draft report "Social Security Numbers (SSN): Internet Resellers Provide Few Full SSNs, But Congress Should Consider Enacting Standards for Truncating SSNs" (GAO-06-495).

We agree that the issue of truncating SSNs (for organizations wishing to do so) would benefit from standardization and we support the recommendation being made to Congress. The results of this review indicate that without a standard truncating method that is widely adhered to, it would be possible for an individual to obtain entire SSNs by purchasing truncated information from one organization that uses the "first five digit" method and purchasing information concerning the same individual from a second organization that uses the "last four digit" method.

Although, as the report accurately states, we do not possess the legal authority to compel organizations to truncate SSNs, or to specify how such truncating should be done, we would be willing to publish information on best practices for truncating SSNs on our Internet site. It would take a few months for this work to be completed.

If you have any questions, please contact Ms. Candace Skurnik, Director, Audit Management and Liaison Staff, at (410) 965-4636.

Sincerely,

Jo Anne B. Barnhart
Jo Anne B. Barnhart

SOCIAL SECURITY ADMINISTRATION   BALTIMORE MD 21235-0001

In: Future of the Internet: Social Networks...
Editors: Rick D. Sullivan and Dominick P. Bartell

Chapter 5

E-GOVERNANCE IN THE PACIFIC ISLANDS: ENTRENCHING GOOD GOVERNANCE AND SUSTAINABLE DEVELOPMENT BY PROMOTING ICT STRATEGIES BASED ON THE RIGHT TO INFORMATION

Indra Jeet Mistry and Charmaine Rodrigues

"We are going through a historic transformation in the way we live, work, communicate and do business. We must do so not passively but as makers of our own destiny. Technology has produced the information age. Now it is up to all of us to build an Information Society."

[1] *(UN,2005).*

1. INTRODUCTION

Many commentators have identified the Pacific Islands countries as being plagued by a diverse range of challenges - political insecurity, ethnic divisions, corruption, economic under-development and social inequality. Making these problems more difficult to tackle is the so-called 'tyranny of distance', the large physical distances and lack of communications infrastructure between the remote communities of the islands and their capitals.

In recent years, Information and Communications Technologies (ICT) have gained prominence for their potential to overcome the tyranny of distance, and in so doing, assist Pacific IslandS Governments to promote their common objectives of good governance and sustainable development, particularly by engaging more effectively with remote and/or rural populations. Indeed, ICTs offer a myriad of ways to connect people in remote locations to their capitals. Although more recent ICTs such as computers are not yet widely entrenched throughout the Pacific, they are gaining traction and could be effectively coupled with older ICTs, such as radios, to great effect.

Recognising the enormous potential for ICTs in the Pacific – if utilised effectively – this paper seeks to raise awareness at this early stage of the importance of ensuring that any ICT strategy is underpinned by a recognition of the 'right to information' and its potential contribution to good governance and sustainability. By purposefully prioritising strategies that entrench the right to information – in particular, proactive disclosure strategies which use ICTs to put more government information into the public domain – ICTs could be used to bridge the governance divide in the Pacific and bring governments closer to their constituents.

2. WHAT IS THE RIGHT TO INFORMATION?

Different terminology has been used - freedom of information, access to information, the right to know - but fundamentally, the concept remains the same. At the heart of the right to information are three key principles:

- the right of the public to request access to information concerning their government's activities.
- the corresponding duty on the government to meet a request, unless specific, defined exemptions apply.
- the additional duty on the government to provide certain key information proactively - even in the absence of a request.

The right to information has been recognised by the United Nations General Assembly as far back as 1946, when it declared *"Freedom of Information is a fundamental human right and the touchstone for all freedoms to which the United Nations is consecrated"* [2]. Soon after, the right to information was given international legal status when it was enshrined in Article 19 of the International Covenant on Civil and Political Rights [3]. Over time, the right to information has been included in a number of regional human rights instruments, including the African Charter on Human and Peoples' Rights, [4] the American Convention on Human Rights [5] and the European Charter of Human Rights [6]. This has placed the right to information firmly within the body of universal human rights law.

In practice, effective implementation of the right to information requires governments to develop legislation, setting out the specific content of the right - such as who people can access information from, how, when and at what cost - and the duties on relevant bodies to provide information - including when they can legitimately refuse to provide information (see Annex 7.1 for details of best practice law-making principles that constitute an effective right to information law).

The Pacific Media and Communications Facility recently released a report – *Opportunities for Media and Communications in the Pacific* – that summarised the state of the right to information in the Pacific region [7]. Most notably, Papua New Guinea's constitution actually specifically entrenches the right to information, [8] although the Government has yet to develop separate right to information legislation. In Fiji, the Constitution explicitly requires the Government to enact legislation on freedom of information [9], but the then SDL Government had indicated its intention to table a Bill [10] as a priority, draft legislation has yet to be released. The only other Pacific Islands country that appears to have made moves to enact a right to information law is the Cook Islands, which earlier in 2005 had completed the drafting of its Official Information Bill which has since been forwarded to the Cabinet for consideration.

3. THE VALUE OF THE RIGHT TO INFORMATION

As noted previously, the right to information imposes a duty on governments to respond to the inquiries of their constituents by providing information on request. It also requires governments to proactively disclose important information to the public, rather than simply conducting its activities in secret. Historically, the right to information was often understood in developing countries as more of an administrative, bureaucratic mechanism. However, experience in many developing countries over the last decade has shown that by empowering citizens with a tool that enables them to scrutinise their government's actions on a day-to-day basis, the right to information is a key mechanism for practically entrenching government transparency, accountability and public participation. Accordingly, an ICT strategy that co-opts the right to information can benefit governance and promote sustainable development in a number of ways.

3.1. Promoting Democratic Governance

The foundation of democracy is an informed constituency that is able to thoughtfully choose its representatives on the basis of the strength of their record and one that is able to hold their government accountable for the policies and decisions it promulgates. The right to information has a crucial role in ensuring that citizens are better informed about the people they are electing and their activities while in government. Democracy is enhanced when people meaningfully engage with their institutions of governance and form their judgments on the basis of facts and evidence, rather than just empty promises and meaningless political slogans. Where people do not have access to information, as has so often been the case in the Pacific Island states, voters will often fall back on ethnic, religious or geographic affiliations when choosing a candidate.

3.2. Supporting Participatory Development and Effective Service Delivery

In terms of promoting sustainable development, much of the failure of development strategies in the Pacific Islands to date is attributable to the fact that, for years, they were designed and implemented in a closed environment - between governments and donors and without the active involvement of the public. However, if governments are obliged to provide information to the public, then citizens can assess for themselves why they have not benefited from government development activities or even from basic service-delivery initiatives in such areas as health care provision and education. Armed with paper-based documentation, the public are in a better position to effectively lobby governments to ensure that services meet their community-specific needs and that money is spent appropriately.

3.3. Supports State-building and Promotes National Stability

Democracy and national stability are enhanced by policies of openness that engender greater public trust in their representatives. Importantly, enhancing people's trust in their government also goes some way to minimising the likelihood of conflict. In post-conflict development situations, the right to information can be especially crucial in the state-building process, such as that underway in Solomon Islands because openness and information-sharing contribute to national stability by establishing a two-way dialogue between citizens and the state, reducing distance between government and people and thereby combating feelings of alienation. Systems that enable people to be part of, and personally scrutinise, decision-making processes reduce citizens' feelings of powerlessness and weakens perceptions of exclusion or unfair advantage of one group over another.

Information empowers citizens and NGOs

The right to information offers a very practical means for individuals, civil society and even parliamentarians to engage more effectively in development activities and ensure that their rights are protected and their legitimate interests promoted. This has been particularly well-illustrated by environmental action groups, which have been very adept at using access to information legislation to expose and discourage anti-green government and/or donor-driven and/or private sector driven programmes.

For example, in 2002 in Uganda, Greenwatch Limited, an environmental NGO, used the open government clause in Article 41 of the Ugandan Constitution to obtain the release of a key document about a controversial dam project that the Ugandan Government and the World Bank had previously declined to release. The Ugandan High Court ordered the release of the document, whose very existence the Ugandan Government had denied during the court proceedings. A subsequent analysis of the document, commissioned by the International Rivers Network assessed that "Ugandans will pay hundreds of millions of dollars in excessive power payments if the World-Bank-financed Bujagali Dam proceeds according to plan." The project was consequently put on hold by the World Bank [11] .

Cracking down on welfare corruption in Delhi, India [12]

In India, poor citizens are given a ration card which entitles them to discounted rice and wheat. This is similar to some of the government subsidy programmes found in the Pacific. In the jurisdiction of Delhi, citizens used their the Right to Information Act to access records held by the private ration shop dealers responsible for administering the programme and found that the shops' owners had been siphoning off rations, while providing false records to the Government Food Department.

Through the right to information applications, citizens could see their ration record held by the Government and compare it to the record in their own ration book. Comparisons showed that the Government's records contained widespread falsifications. The campaign to open up the registers of ration shop dealers and the Government took two years and met with stiff resistance from Government and the private dealers – including death threats and physical attacks. However, at the end of the campaign, the Delhi Government has now agreed to proactively open up ration registers every month to enable citizens to regularly check their ration records. It is expected that this will substantially reduce economic losses to the Government from misdirected welfare and ensure proper benefits to the poorest in the community.

3.4. Exposes Corruption

If implemented effectively, the right to information can act as a powerful deterrent of government corruption. In many Pacific states, corruption has eaten away at civil society's trust in the state's political and legal system and has greatly hindered economic development. In particular, corruption can destroy efforts at poverty reduction, creating a vicious circle where funds are often skimmed off by the state, which, in turn, deters crucial foreign investment and growth. However, an effective right to information regime can make it much harder for government officials to cover up their corrupt practices at all levels of government, and also works to expose poor policymaking. Notably, in 2004 of the ten countries scoring best in Transparency International's annual Corruption Perceptions Index, no fewer than eight had effective legislation enabling the public to see government files. In contrast, of the ten countries perceived to be the worst in terms of corruption, only one had access to information legislation in place.

3.5. Supports Equitable Economic Development and Investment-friendly Growth

In recent years, Pacific Islands states have struggled to attract significant levels of foreign investment in order to accelerate economic growth and development. The benefits from economic development have often not been equitably enjoyed by all citizens. However, a transparent government that is committed to right to information is obliged to provide good-quality economic and social data proactively, which will better inform government economic policy and decision-making. Openness about licensing regimes, awareness of concessions and permits and more accountable government decision-making processes will also bolster private and foreign investor confidence in the economy, encouraging long-term private investment and thereby boosting growth. By empowering smaller stakeholders to more effectively participate in the economy, the right to information can also help to ensure the economy grows in an equitable manner.

4. Pacific ICTs Strategies Should Prioritise the Promotion of the Right to Information

Work on the right to information in the Pacific region has been limited, but already there has been significant interest in utilising ICTs at last to offer to promote good governance and more sustainable and participatory development. ICTs offer an opportunity to finally address the 'tyranny of distance' which has plagued Pacific governance and development strategies for so long, by making geographic constraints redundant, providing remote communities with the means to access the 'information highway' to their capitals and the world beyond.

Nevertheless, although ICTs provide important tools to address the tyranny of distance, if they are to be utilised to maximum advantage in terms of promoting sustainable development and good governance in the Pacific Islands, it is strongly recommended that specific attention be paid at these early stages to ensuring that Pacific ICT strategies properly incorporate the promotion of the right to information.

4.1. A 'Proactive Disclosure' Approach

Ideally, the right to information should be underpinned by comprehensive legislation. However, in the Pacific, legislative processes have proven to be slow and unreliable – Fiji's Constitution *requires* the enactment of legislation but governments have yet to pass a right to information law seven years since the Constitution came into effect! Notably though, this does not mean that ICTs cannot still be used to entrench the right to information in practice to ensure that the benefits of the right can still be enjoyed by citizens of the Pacific Islands.

ICT strategies can be harnessed to prioritise the concept of 'proactive disclosure' – getting governments to routinely disclose and disseminate information into the public domain even in the absence of a specific request for it. Significantly, this emphasis on proactive disclosure would actually draw upon what is becoming an international trend. Early right to information laws – including the access laws passed by New Zealand [13] and Australia [14] in 1982 – focused heavily on the public's right to request specific information from the Government. However, over the last decade, newer right to information statutes have put greater emphasis on the principle of proactive disclosure, in recognition of the fact that if the right is really to be of assistance to the poorest and most disadvantaged, it should place as little burden on them as possible, and instead should actively provide them with information in order to facilitate their re-engagement with government.

Notably, if implemented with public participation objectives in mind, proactive disclosure activities can be a very cheap but effective way of interacting with citizens. The most notable examples of such initiatives are contained in the very strong Mexican *Transparency Law* passed in 2000, and the new Indian *Right to Information Act* passed in 2005 which both include proactive disclosure requirements requiring all public authorities to publish and disseminate more than 15 different categories of information, from basic information about departmental organisation structures and services, to detailed information about public tenders and contracts awarded and recipients of government concessions and subsidies. Most of this information is to be published on government websites – a clear interface between ICTs and the right to information with a view to promoting better governance.

5. PRACTICAL IDEAS FOR IMPLEMENTATION

In the Pacific, the right to information has been so overlooked as a governance and development strategy that opportunities exist at all levels to incorporate the right into ICT activities and in so doing, to ensure that the ICTs activities undertaken by Pacific governments are the most appropriate and will most effectively promote good governance and participatory development in both the short and long term. A few areas for work that could be undertaken as an immediate priority are discussed below.

5.1. Developing a Coordinated Information and ICTs Policy

As noted above, although a right to information law is unlikely to be passed in the short term in most Pacific Islands countries, nonetheless, government information and ICT policies that promote proactive disclosure could offer a very good start towards this goal. Information and ICT policies can prioritise proactive disclosure as a means of providing a practical response to the public participation problems identified across the region. It is notable in this regard that many ICT strategies focus on using ICTs to promote better service delivery. While this is an important goal in and of itself, it overlooks the potential for ICT actually to promote broader democratic values and aims – such as transparency, accountability and participation – and instead reduces ICT to promoting only operational ends. This approach should be reconsidered.

Information and ICT policies need more explicitly to recognise the potential for ICT to promote public participation in governance and development. The public needs to be recognised as not mere passive receivers of government services through web-based mechanisms, but as active partners who should be engaged and supported to interact with the government, most notably through ICT-based dissemination of relevant government information.

Refocusing Pacific ICT and information disclosure policies

In the last year, CHRI has been strongly encouraging many national governments in the Pacific as well as the Pacific Forum Secretariat to promote greater information disclosure, at a minimum via the internet. It was recognised early on that statutory-based information regimes were probably not viable in the short term, but that nonetheless, the people of the Pacific were entitled to have access to more information about who governed them, how and to what end. The Forum Secretariat has been encouraged to include the promotion of the right to information as an objective of the new Pacific Plan. It has also been encouraged to develop a Forum Secretariat Information Disclosure Policy, which would ensure that all except the most sensitive Secretariat documents were published on the Secretariat's website, for download and dissemination within Forum member states.

ICTs can provide effective strategies to deliver government information to remote communities in Pacific countries. However, ICTs have become synonymous with Internet and digital technology while many parts of the Pacific Islands still do not have access to electricity and a telephone line, let alone to a computer. In this context, it is important that ICT and Information Policies recognise the obvious – that ICTs in the Pacific include not only new computer-based e-governance initiatives, but also capture older technologies such as radios. Ideally, these technologies should *all* be utilised, as resources, capacity and geography permit.

In the Pacific, this would necessitate that any ICT strategy clearly recognise the important role that battery powered radios have played in terms of connecting up remote communities. To utilise current radio systems more effectively though, consideration should be given to using the radio more regularly as a cheap but effective method for communicating information on government policies and activities to the public. Special radio programmes updating communities on the development and implementation of key government policies and activities could be run by the government. For example, radio programmes could be developed to update communities on when provincial health service grants have been paid, so that members of the community can then know to question their service providers on what they are doing with the funds. Governments could also use the radio to announce national policies, such as the budget, and explain their specific objectives and likely impacts in simple terms.

Any Information and ICT Policies could also consider the development of community information hubs based around current ICT infrastructure. For example, many rural health clinics contain a radio that is used to link up the clinic with headquarters and other health clinics so that they can get advice from specialists and/or check diagnoses and/or for training purposes. However, health clinics are a community resource more generally and their radios could be considered similarly. Health clinic radios, their noticeboards and possibly even their computer infrastructure, could be utilised as information dissemination points for all members of the community (subject of course, to the needs of the health clinic).

Even new ICT infrastructure should be recognised for the potential as information dissemination points. For example, in Solomon Islands the new PFNet system, which is focused on setting up computer-based "information kiosks" in rural areas, could be developed to ensure that those information kiosks allow quick access to key government information, by including sites in the list of favourites and/or by developing a new site – which could be used as the default page when the browser opens – that collects together all key government links.

5.2. Promoting More Participatory Governance

At the moment, there is a dearth of useful, up-to-date information in the Pacific about Pacific governments. Only a very small number of Pacific governments have websites that capture information about all government departments. More commonly, one or two departments may simply set up websites on their own, highlighting only their own work, and even in these cases, the websites are often not updated.

This makes it extremely difficult even for citizens in the capital city to engage effectively with their government. Identifying who is responsible for what service or policy can be extremely difficult, while sending a fax or email to that person can be near impossible. Simple forms can be difficult to get hold of, unless a citizen is prepared to actually attend at the relevant department, while it can also be very difficult for people to find out about services and benefits to which they are entitled.

In this environment, ICT clearly need to be harnessed more effectively to promote simple, cheap and user-friendly access to governance information. Ideally, governments should prioritise as one of their first ICT initiatives the collection and dissemination of this information (see the box below for a list of the most essential information). The information could be relatively cheaply published on the internet, although obviously to ensure greater outreach, consideration could be given to using other ICT technologies such as radio, TV and newspapers.

Notably, even if the new governance information is initially provided on the internet, the usefulness of this form of dissemination should not be underestimated simply because many Pacific communities do not currently have good access to the internet.

Any ICTs and Information Policies should recognise the importance of developing outreach partnerships so that new technologies can be utilised by existing networks for maximum effect.

In the Pacific, the most obvious strategy in this regard is supporting – formally or informally – church and civil society groups who already have strong networks in rural areas to access the information on the web so that they can then disseminate it through their own channels.

Extending the Pacific Portal?

The University of the South Pacific and Institute of Justice and Applied Legal Studies (IJALS) are jointly developing an e-Governance portal. Known as the Pacific Portal, it is understood that the site will focus largely on ensuring the statistical and other data in specific sectors, namely education and health, are comprehensively and consistently collected and placed on the internet.

Notably, however, there is huge potential to extend the scope of the site to include basic government information from across the Pacific, and in so doing, to bring communities across and within Pacific Island Countries closer together. In 2004, CHRI suggested that the Portal could be broadened to include information on the duties, responsibilities and services provided by all levels of government and its departments (see Appendix 7.2 for more). It was suggested for example, that the Portal could include the following information (which of course, ideally should already be collected and published by member country governments):

- *A list of every government department, including:*
  - *the name and contact details of the responsible Minister*
  - *the name and contact details of the CEO/Permanent Secretary*
  - *the physical address and mailing of the department*
  - *a statement of the aims and objectives of the department*
  - *a chart/list setting out the organisational structure of the department and/or a directory of the departments public servants, at least to the level of Section Head (or the equivalent), including the total number of staff in the department broken down by levels and the pay scales applicable to each level*
  - *the services offered, schemes run, subsidy programmes implemented by the department; including any relevant copies of all policies, guidelines and forms*
  - *a description of the powers and duties of senior officers and the procedure to be followed in making decisions*
  - *an explanation of any departmental complaints mechanisms*
  - *a register of the types/categories of information/records the department holds and publishes and the procedure to be followed in obtaining information*
- *If resources permit, a list of every major public authority, such as the Reserve Bank, Office of the Ombudsman, Pubic Service Commission, etc.*
- *The Government's annual budget, including a breakdown by department and further breakdowns by line item if possible*
- *Regularly updated reports (quarterly if possible) about the disbursement of the Budget;*
- *The results of any Government audits and corresponding departmental explanations;*
- *Information about any Government inquiries/commissions, including copies of submissions and draft and final reports;*
- *Mechanisms for citizen participation, where they exist.*

5.3. Supporting More Effective (Decentralised) Service Delivery

A proactive disclosure-focused ICT strategy can also be targeted towards promoting more effective service delivery, in particular by facilitating more informed participation by communities in the operation and oversight of community services. In this context, it is notable that a number of Pacific Islands countries have chosen to pursue devolution or decentralisation strategies, with a view to bringing governance closer to communities. However, these strategies have been fraught with difficulty. Local level corruption is a major problem and effective public participation is still lacking. Making things worse, there is still often considerable confusion, not only among communities but also within government, regarding which department or level of government is responsible for providing what public services. This weakens accountability mechanisms.

As noted previously, however, at a more general level, ICTs could be used as an effective mechanism for proactively disclosing more information about what the government does and who does it. In terms of service delivery specifically, this means that ICTs could be used to disseminate information about how much money is to be spent on a community, what the money is to be spent on, over what period of time and by which department(s). Armed with this information, the public can start demanding more effective service-delivery from the various levels of governments. In fact, ICTs can provide a two-way channel of communication between governments and remote communities that empowers local citizens with real influence over decisions that affect their lives.

Using ICT to promote more effective delivery of education and health services

In the Pacific, health and education services have commonly been developed to local government bodies to implement. Monthly or quarterly grants are sent from the central government to the local/provincial government, which is supposed to administer it and report back on its expenditure. Unfortunately, however, it has been witnessed that that such arrangements have resulted in considerable leakage of funds, with little accountability as the various levels of government blame each other. The public has difficulty knowing who to hold responsible because they have little or no information on when the grant was to be paid, whether it was paid, how much it was for and what services they were supposed to get as a result. All they know is that government – at some or all levels – is not performing.

Using ICTs to proactively disclose more information to communities about government expenditure could be a key way of addressing this problem, as a case study from Uganda shows. In that country, despite the increases in education expenditure during the 1990s, a five-year survey found that 87% of all funds meant for primary schools throughout the country were being pocketed by bureaucrats. Shocked by these findings, the national government began to publicise details of monthly transfers of grants to districts through newspapers and the radio in order to curb the siphoning of funds. Primary schools were also required to display public notices on receipt of their funds.

Empowered with this information, parents could monitor the educational grant programme and make the local government accountable for education provision. In five years, the diversion of funds dropped massively from 80% to 20%, while enrolment more than doubled from 3.6 million to 6.9 million children. In this way, proactive disclosure of information concerning education funding, though a simple and inexpensive strategy, enforced greater local government accountability and ensured the proper use of taxpayers' money [15]

CONCLUSION

ICTs now provide an opportunity to overcome physical boundaries and have been hailed as a solution to the Pacific Islands' chronic problems of governance. What they must do, as a matter of priority, however, is to deliver the immediate benefits of the right to information to Pacific citizens – most easily by helping governments to publish and disseminate a wide range of key government information proactively.

To date, in the Pacific Islands, the paucity of physical infrastructure between governments and their remote communities has made the delivery of government information near impossible. However, if ICT and right to information strategies are developed in a coordinated fashion, together they can be harnessed to open up channels of communication between Pacific Islands governments and their disparate populations, thereby not only narrowing the physical tyranny of distance but also bridging the divide that exists in terms of public trust, transparency and accountability.

REFERENCES

[1]   UN Secretary General, Kofi Annan, *Message to World Summit on Information Society*, Tunis 2005. http://www.itu.int/wsis/messages/annan.html.

[2]   See UN General Assembly Resolution 59(1), 65th Plenary Meeting, 1946.

[3]   This states: "Everyone has the right to freedom of opinion and expression; this right includes freedom to hold opinions without interference and *to seek, receive and impart information and ideas through any media and regardless of frontiers.*" [emphasis added]

[4]   See *A Right to Information*, Article 9(1), *African Charter on Human and Peoples' Rights*, OAU Doc. CAB/LEG/67/3 rev. 5, 21 I.L.M. 58 (1982), 27 June 1981.

[5]   See *A Right to Information,* Article 13(1), *American Convention on Human Rights*, 1969, Costa Rica, OAS Treaty Series No. 36, 1144 U.N.T.S. 123.

[6]   See *A Right to Information,* Article 11(1), *Charter of Fundamental Rights of the European Union*, 2000, Nice, Official Journal of the European Communities, C 364/1.

[7]   For more information see the Pacific Media and Communications Facility website at http://pmcf.muprivate.edu.au/index.php?id=798. See also CHRI's website at www.humanrightsinitaitive.org (click on Right to Information, International, Member States Laws and Papers).

[8]   Article 51 of the Constitution explicitly recognises the right of reasonable access to official documents, subject only to the need for such secrecy as is reasonably justifiable in a democratic society.

[9]   Article 30(1) of the Constitution includes the freedom to seek, receive and impart information and ideas as part of the right to freedom of expression. Article 174 explicitly requires that Parliament should enact a law to give members of the public rights of access to official documents of the Government and its agencies, as soon as practicable after the commencement of the Constitution.

[10]   Minister for Information Marieta Rigamoto announced at the end of September that the Government was drafting a Freedom of Information Bill which, along with the Public

Records Act, would give citizens wider access to information held by the Government and its agencies. See http://www.fijitimes.com/story.aspx?id=29425 for more details.

[11] *Ugandan Judge Orders Release of Key Document on Bujagali Dam*, 22 November 2002, http://www.freedominfo.org/ifti1102.htm#1 as at 22 July 2003.

[12] *Look, How Fair Price Shops Clean Up Act*, Indian Express, February 24, 2005. See http://cities.expressindia.com/ fullstory.php?newsid=118913.

[13] See   http://www.legislation.govt.nz/libraries/contents/om_isapi.dll?clientID=126036 5289&hitsperheading=on&infobase=pal_statutes.nfo&jump=a1982- 156&softpage=DOC for a copy of the law.

[14] See http://www.scaleplus.law.gov.au/html/pasteact/0/58/top.htm for a copy of the law.

[15] World Bank (2001), *World Development Report 2000-01: Attacking Poverty*, Oxford University Press, New York

[16] CHRI (2003), Open Sesame: Looking for the Right to Information in the Commonwealth" Chapter 2. See www.humanrightsinitiative.org.

[17] Fiji - www.fiji.gov.fj, Nauru - www.un.int/nauru; PNG - http://www.pngonline.gov.pg; Samoa - www.samoa.ws; Solomon Islands - www.commerce.gov.sb; Tonga - www.pmo.gov.to; Tuvalu - www.tuvaluislands.com; Vanuatu - www.vanuatugovernment.gov.vu.

APPENDIX 7.1.
KEY PRINCIPLES UNDERPINNING A GOOD RIGHT TO INFORMATION LAW

Governments can grant access to information through executive orders, administrative guidelines and some have even entrenched Right to Information in their constitutions. Nevertheless, to ensure consistent application of simple, clear, agreed processes, Right to Information legislation is still essential. A number of basic principles that have evolved over time that should be enshrined in all access laws, most of which have been consistently endorsed by the UN, the Commonwealth, the Organisation of American States and the African Union [16]:

1. *Maximum Disclosure*: The value of access to information legislation comes from its importance in establishing a framework of open governance. This means that there should be a presumption in favour of access - bodies covered by the Act have an obligation to disclose information and every member of the public has a corresponding right to receive information. Any person should be able to access information, whether a citizen or not. All arms of Government should be covered – executive, judiciary and legislature – and private bodies should also be subject to the law, at least where the information requested affects people's rights. Bodies covered by the Act should also be required to publish and disseminate documents of general relevance proactively to the public, for example, on their structure, norms and functioning, the documents they hold, their finances, activities, any opportunities for consultation and the content of decisions/policies affecting the public.

2. *Minimum Exceptions:* The key aim of any exceptions to disclosure should be to protect and promote the public interest. The law should therefore not permit non-disclosure in order to simply protect government from embarrassment or the exposure of wrongdoing. In line with the commitment to maximum disclosure, exemptions should be kept to an absolute minimum and should be narrowly drawn. Blanket exemptions for specific positions (eg. President) or bodies (eg. the Armed Services) should not be permitted. Even where exemptions are allowed, they should still ALL be subject to a blanket "public interest override", whereby a document which is presumed exempt under the Act should still be disclosed if the public interest in the specific case requires it.

3. *Simple Access Procedures:* A key test of an access law's effectiveness is the ease, inexpensiveness and promptness with which people seeking information are able to obtain it. The law should include clear and uncomplicated procedures that ensure quick responses at affordable fees. Applications should be simple and devised to ensure that the illiterate and/or impecunious are not in practice barred from utilising the law. Any fees that are imposed for gaining access should also not be so high as to deter potential applicants and should be limited only to cost recovery. The law should provide strict time limits for processing requests.

   All public bodies should be required to establish open, accessible internal systems for ensuring the public's right to receive information. Provisions should be included in the law which require that appropriate record-keeping and record management systems are in place to ensure the effective implementation of the law.

4. *Independent Appeals Mechanisms:* Effective enforcement provisions ensure the success of access legislation. Powerful independent and impartial bodies must be given the mandate to review refusals to disclose information and compel release. While internal appeals provide an inexpensive first opportunity for review of a decision, oversight by an umpire independent of government pressure is a major safeguard against administrative lethargy, indifference or intransigence and is particularly welcome where court-based remedies are slow, costly and uncertain. The fear of independent scrutiny ensures that exemption clauses are interpreted responsibly and citizens' requests are not unnecessarily obstructed. While the courts satisfy the first criteria of independence, they are notoriously slow and can be difficult to access for the common person.

   As such, in many jurisdictions, special independent oversight bodies have been set up to decide complaints of non-disclosure. They have been found to be a cheaper, more efficient alternative to courts and enjoy public confidence when they are robustly independent, well-funded and procedurally simple. Best practice supports the establishment of a dedicated Information Commission with a mandate to review refusals to disclose information, compel release and impose sanctions for non-compliance.

5. *Penalties:* The powers of oversight bodies should include a power to impose penalties. Without an option for sanctions, such as fines for delay or even imprisonment for wilful destruction of documents, there is no incentive for bodies subject to the Act to comply with its terms, as they will be aware that the worst that can happen is simply that they may eventually be required to disclose information.

In the first instance, it is important to detail what activities will be considered offences under the Act, for example, unreasonable delay or withholding of information, knowing provision of incorrect information, concealment or falsification of records, wilful destruction of records without lawful authority, obstruction of the work of any public body under the Act and/or non-compliance with the Information Commissioner's orders. Once the offences are detailed, sanctions need to be available to punish the commission of offences. International best practice demonstrates that punishment for serious offences can include imprisonment, as well as substantial fines. Notably, fines need to be sufficiently large to act as a serious disincentive to bad behaviour.

6. *Monitoring and Promotion of Open Governance:* Many laws now include specific provisions empowering a specific body, such as an existing National Human Rights Commission or Ombudsman, or a newly-created Information Commissioner, to monitor and support the implementation of the Act. These bodies are often be empowered to develop Codes of Practice or Guidelines and are also usually required to submit annual reports to Parliament. Although not commonly included in early forms of right to information legislation, it is increasingly common to include provisions in the law itself mandating a body to promote the Act and the concept of open governance. Such provisions often specifically require that the government ensure that programmes are undertaken to educate the public and the officials responsible for administering the Act.

APPENDIX 7.2.

A Proposal to Extend the Pacific Governance Portal

1. The Pacific Governance Portal currently being developed by IJALS provides an exciting opportunity to extend the right to information to the people of the Pacific, despite the absence of a single right to information law in any Pacific country.

Value of the Right to Information

2. *It strengthens democracy*: The foundation of democracy is an informed constituency that is able to thoughtfully choose its representatives on the basis of the strength of their record and that is able to hold their government accountable for the policies and decisions it promulgates. The right to information has a crucial role in ensuring that citizens are better informed about the people they are electing and their activities while in government. Democracy is enhanced when people meaningfully engage with their institutions of governance and form their judgments on the basis of facts and evidence, rather than just empty promises and meaningless political slogans.

3. *It supports participatory development*: Much of the failure of development strategies to date is attributable to the fact that, for years, they were designed and implemented in a closed environment - between governments and donors and without the

involvement of *people*. If governments are obligated to provide information, people can be empowered to more meaningfully determine their own development destinies. They can assess why development strategies have gone askew and press for changes to put development back on track.

4. *It is a proven anti-corruption tool*: In 2003, of the ten countries scoring best in Transparency International's annual Corruption Perceptions Index, no fewer than eight had effective legislation enabling the public to see government files. In contrast, of the ten countries perceived to be the worst in terms of corruption, not even one had a functioning access to information regime. The right to information increases transparency by opening up public and private decision-making processes to scrutiny.

5. *It supports economic development*: The right to information provides crucial support to the market-friendly, good governance principles of transparency and accountability. Markets, like governments, do not function well in secret. Openness encourages a political and economic environment more conducive to the free market tenets of 'perfect information' and 'perfect competition'. In turn, this results in stronger growth, not least because it encourages greater investor confidence. Economic equity is also conditional upon freely accessible information because a *right* to information ensures that information itself does not become just another commodity, to be cornered by a few for their sole benefit.

6. *It helps to reduce conflict*: Democracy and national stability are enhanced by policies of openness, which engender greater public trust in their representatives. Importantly, enhancing people's trust in their government goes some way to minimising the likelihood of conflict. Openness and information-sharing contribute to national stability by establishing a two-way dialogue between citizens and the state, reducing distance between government and people and thereby combating feelings of alienation. Systems that enable people to be part of, and personally scrutinise, decision-making processes reduce citizens' feelings of powerlessness and weaken perceptions of exclusion or unfair advantage of one group over another.

The Pacific Governance Portal and the Right to Information

7. The right to information is most commonly understood to refer to the right of the public to obtain access to government information upon request and the concomitant duty of government to ensure that systems are in place to facilitate the efficient exercise of that right. However, over the last decade, the right to information has come to more broadly capture the additional duty of governments to proactively provide the public with important and/or routine information, for which specific requests need not be made.

8. It is the second feature of the right to information – referred to for quick reference as "proactive disclosure" – that the PGP can be developed to address, even in the absence of right to information legislation. The PGP is currently focusing largely on ensuring the statistical and other data in specific sectors, namely education and health, is comprehensively and consistently collected and placed on the internet. However, considering the state of many Government websites in the Pacific, CHRI

encouraged IJALS to consider adding an additional section to the PGP - as a matter of priority - which would enable the public to access basic government information.

9. Currently, all Commonwealth Pacific member states except Kiribati have Government Websites [17]. However, their content varies considerably. Solomon Islands for example, only has websites for the PMO and the Commerce Department. Conversely, Tonga has a relatively comprehensive website, with links to all Ministries, including contact details, summaries of their responsibilities and links to some key documents.

10. It would be a major contribution to the Pacific to equalise the access of Pacific people to information about their Governments by ensuring that all Pacific Government websites contain a minimum amount of information. The PGP provides a unique and timely mechanisms for achieving this objective. In keeping with international best practice in terms of proactive disclosure, the new Government Information Section of the PGP could include the following information: .

- *A list of every government department, including:*
  - *The name and contact details of the responsible Minister*
  - *The name and contact details of the CEO/Permanent Secretary*
  - *The physical address and mailing of the department*
  - *A statement of the aims and objectives of the department;*
  - *A chart/list setting out the organisatonal structure of the department and/or a directory of the departments public servants, at least to the level of Section Head (or the equivalent), including the total number of staff in the department broken down by levels and the pay scales applicable to each level*
  - *The services offered, schemes run, subsidy programmes implemented by the department; including any relevant copies of all policies, guidelines and forms;*
  - *A description of the powers and duties of senior officers and the procedure to be followed in making decisions;*
  - *An explanation of any departmental complaints mechanisms;*
  - *A register of the types/categories of information/records the department holds and publishes and the procedure to be followed in obtaining information;*
- *If resources permit, a list of every major public authority, such as the Reserve Bank, Office of the Ombudsman, Pubic Service Commission, etc.*
- *The Government's annual budget, including a breakdown by department and further breakdowns by line item if possible*
- *Regularly updated reports (quarterly if possible) about the disbursement of the Budget;*
- *The results of any Government audits and corresponding departmental explanations;*
- *Information about any Government inquiries/commissions, including copies of submissions and draft and final reports;*
- *Mechanisms for citizen participation, where they exist.*

11. Although this information will initially require a considerable amount of time to collect properly, the benefits in terms of more transparent and accessible governance in the Pacific would be enormous. Not only local citizens, but also people from throughout the region and even further afield would be able to interact more easily with Pacific governments. This could have particular benefits in terms of foreign investment. Once the initial collection was completed, updating the information would require less effort. Best practice requires updating every 6 months, but common practice sees government providing updates only annually or biennially.

In: Future of the Internet: Social Networks...   ISBN: 978-1-61209-597-4
Editors: Rick D. Sullivan and Dominick P. Bartell   ©2011 Nova Science Publishers, Inc.

Chapter 6

EXPLORING THE USE OF WEB 2.0 TOOLS TO SUPPORT COLLABORATIVE LEARNING

Qiyun Wang and *Huay Lit Woo*

Learning Sciences and Technologies Academic Group,
National Institute of Education, Nanyang Technological University,
1 Nanyang Walk, Singapore 637616

ABSTRACT

The ability to collaborate is becoming more and more important in today's world in which tasks are getting more and more interdisciplinary and complicated to accomplish. It is therefore essential to prepare students on collaborative tasks while they are in schools so that they can become competent team workers when they enter the workforce. This paper illustrates how various web 2.0 tools have been used to support collaborative learning. The web 2.0 tools presented in this paper include Weblog, Wiki, Google Docs, Yahoo group, and Facebook. The affordances of these tools for collaborative learning together with examples of using the tools to support teachers and students in collaborative learning processes are also described.

INTRODUCTION

Collaboration is an essential competency in the current knowledge society. In the new information age, work becomes more knowledge-based, interdisciplinary and complicated. It is difficult for an individual to complete a sophisticated task without the help of others. The ability to work collaboratively hence becomes highly valued in the present workplace (Barron, 2000). Collaborative learning has the promise of active construction of knowledge, enhanced problem articulation and promotion for social interaction (cf. Haythornthwaite, 2006). It has also demonstrated better learning outcomes than individual work in numerous

---
* Tel: +65 6790 3267; Fax: +65 6896 8038; Qiyun.wang@nie.edu.sg

studies (e.g. Barron, 2000; Lipponen, Hakkarainen, and Paavola, 2004; Neo, 2003). With these advantages, it is therefore crucial to prepare students to work collaboratively while they are in schools. This paper describes how web 2.0 tools have been used to support collaborative learning. The web 2.0 tools presented in this paper include Weblog, Wiki, Google Docs, Yahoo group and Facebook.

WEB 2.0 TOOLS

Tools that can be used to support CSCL must fulfill the need to provide affordances that are conducive to collaborative learning. They must have facilities to engage students in group activities. Most web 2.0 tools possess the above requirements and are useful for learning. Web 2.0 is the second-generation of web-based tools that enhance sharing, interaction and collaboration among its users. The first-generation of web technologies (or called web 1.0) is primarily meant to deliver information from the web server to users. It mainly supports one-way information delivery only. The web 2.0 technology, however, enables two-way communication. The users are not only information consumers, but also information contributors. They can create and upload new information to the web server. Also, they can modify information published. Web 2.0 web pages such as iGoogle are more likely to be multimedia programs, in which users can create new tabs, add new objects, or drag-and-drop objects to any position on a web page. The advent of web 2.0 technology has made many web-based applications more interactive than before. Weblog, Wiki, Google Docs, Yahoo groups, and Facebook are representative examples of Web 2.0 tools. The subsequent sections present how these web 2.0 tools can be effectively used to support collaborative learning.

Weblog

A Weblog is a platform in cyberspace to allow one to publish diaries or journals. In education, Weblogs can be used by students to write reflections or share resources with fellow students. It can also be used by teachers to broadcast course announcements, display courses resources or collect feedback from students (Wang and Woo, 2008a). Therefore, Weblogs can provide two types of interaction, the student-to-student and teacher-to-student interactions.

Sometimes it is necessary to create a learning environment that involves both types of interaction depending on the needs of leaning. For example, in a guided collaborative learning environment in which students work in groups to accomplish tasks and at the same time receive constant guidance from the teacher in the form of scaffolding. This would require a common working space in which both the teacher and students can interact with one another. This is made possible by using Weblogs in eBlogger (http://www.blogger.com). The group of students can collectively own an eBlogger blog space but they must authorize the teacher as an additional author of their blogs so that both the teacher and students can co-publish the blogs and communicate freely through the use of the "comment" function in the blog.

Below uses a real classroom example to illustrate how student-student-teacher interaction could take place in an environment using Weblog as the interactive tool.

A Classroom Example of Using Weblog

A class of student teachers from an elective course of Multimedia Instructional Design took part in activities which use eBlogger as a collaborative tool. The student teachers were second year university students who were pursuing a Diploma in Education in National Institute of Education (NIE) of Singapore (Wang and Woo, 2008b).

Two strategies were deployed to provide the collaborative learning, they were: group collaboration and whole class sharing. First, for group collaboration, the class was divided into groups of two or three students where each group was to work on a final project that aimed at developing a multimedia-based learning package for use in a primary school setting. During the project development stage, the group members used the Weblog to share, negotiate and discuss design ideas with fellow members as well as with the tutor. It is noted that student teachers benefit from two affordances provided by the Weblog. First, the Weblog affords group members the opportunity to share resources and ideas conveniently and a means to communicate easily with each other. Second, it affords the tutor to track the developmental progress of the group and the members' individual contributions. In other words, a Weblog allows collaboration to take place within a conducive environment and a permanent record of the collaboration process. Figure 1 shows a screen shot of the group's Weblog.

Next, for class sharing, the purpose is to extend the interaction from groups to the whole class. This is done by creating a blog-based "bulletin board" for members of the entire class to exchange ideas, ask questions and seek help from each other. Each group was asked to put up their group's Weblog URL in this shared space so that other members of the class could find out more about what their peers were doing and how they were progressing. The ability to obtain up-to-date information helps the class to keep in pace with each other and motivate any straggler to catch up with the rest. The groups were also required to post their final project proposal to the class shared corner and to invite others for comments and opinions.

The feedback gathered could help them improve on the quality of their projects and detect any conceptual problems which otherwise were difficult to be noted for an individual. Student teachers were happy with using Weblog for the group collaboration and whole class sharing, reporting that the Weblog activities helped and facilitated the completion of their final projects. Figure 2 below shows how each group used the class shared corner to show their project proposal and collect feedback from the others.

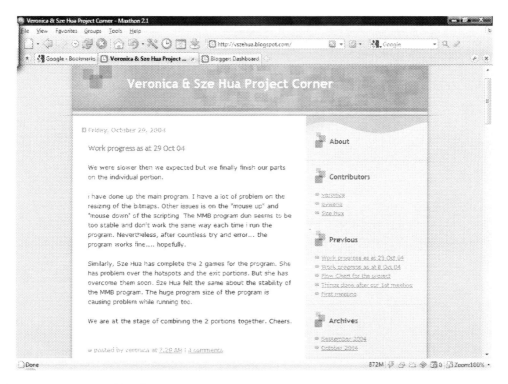

Figure 1. A group's Weblog for the final project.

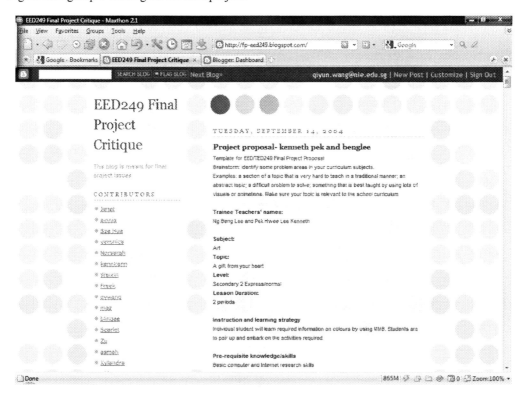

Figure 2. Use Weblog to show project proposals and collect feedback.

Wiki

A Wiki is a web page or a collection of web pages designed to enable anyone who has editing rights to modify or co-write a piece of document in an online environment. The Wiki used in this paper is PbWiki (http://pbwiki.com). Unlike the Weblog whose ownership belongs to the author, the Wiki believes that "the whole is more than the sum of the parts" and as such, it allows all authorized members to co-own the Wiki site. A Wiki has another advantage; it allows volunteering readers to make contribution to the content by requesting for co-authorship. A famous Wiki site which makes use of such mechanism to develop a knowledge base is the Wikipedia (http://www.wikipedia.org). Wikipedia is a product of many people's contribution to specific areas. What makes Wikipedia very popular is its ability to attract experts of all fields to come together to co-write a certain topic. Because of this, Wikipedia is able to present a topic or subject from different perspectives, which is why it is suffixed with the word "pedia" - a short form of the term "encyclopedia".

Like any other tools, a Wiki is also limited by its constraints. One of these constraints is that it cannot automatically screen away untrue or unreliable contributions. To get around with this, human intervention is often required. Fortunately, in education, a Wiki is normally incorporated in a closed learning environment with a teacher or tutor as a moderator. This is because in education, accessing to the right information is as crucial as understanding the information. Hence, learning is not entirely unguided but always assisted by the scaffolding of a teacher or tutor. This reduces the possibility of learning improper information.

A Classroom Example of Using Wiki

Presented below is an example of how the Wiki technology was applied to facilitate group collaborative activities. The Wiki activities described here were attended by a class of post-graduate student teachers doing a course on "Designing effective learning environment". As part of the course requirements, the student teachers were formed into groups of four. The groups were instructed to work collaboratively to see how they could modify and improve on an existing set of rubrics to make it suitable for assessing learning environments. The existing rubrics consisted of four criterion categories: usability, content, educational value and vividness. The rubrics were originally designed for assessing the quality of an educational website, not for a learning environment. So group members had to put on their thinking caps to find ways to improve the rubrics using their understanding of what constitute a good ICT-based learning environment. Figure 3 shows a partial screen shot of a group's members interactivity recorded by the Wiki page.

It records 23 revisions done over a four-day period. The Wiki also identifies the contributors and shows the content which the contributors had modified. Figure 4 shows the changes made by two of the group members on the same day. To extend the interactivity among the learners, the student teachers were also required to visit other group's Wikis and provide comments when necessary.

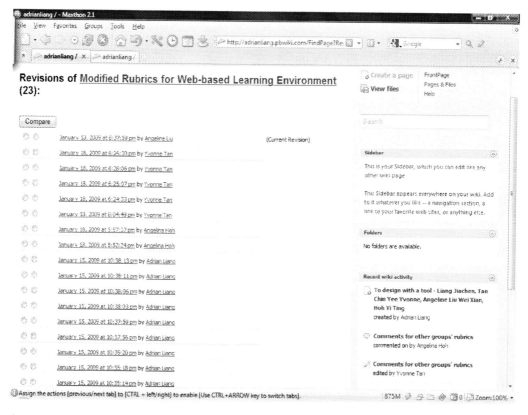

Figure 3. The history page of a group's Wiki.

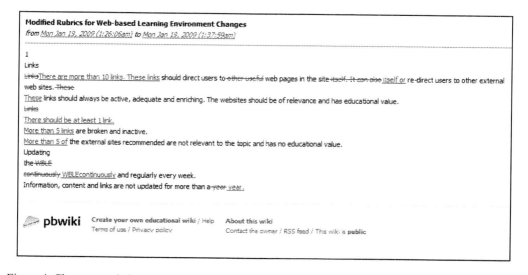

Figure 4. Changes made by two group members shown on the Wiki page.

A Wiki is not without pitfalls. This is because pedagogically, a Wiki can provide a useful means for collaborative work to take place. But technically, a Wiki cannot provide synchronous interaction within the same time frame. This means that a Wiki can only allow one person at a time to do the editing, any other members trying to co-edit simultaneously will be denied of the rights for their contribution. In other words, all co-editors of the same content will have to take turn to write and the turn-taking process is based on the rule of "first-come-first-serve". This implies that a Wiki will need the collaboration to take place somewhat linearly, but nonetheless, it gives the co-authors longer time for reflection and lesser chance to be impulsive in their writing. This suggests that perhaps Wiki is a better tool for reflective learning when pace of learning is not an issue but quality of work is more crucial.

Google Docs

Google Docs is another web-based tool that works similarly with a Wiki. But unlike the Wikis which limit the editing process to "one-at-a-time", a Google Docs can allow several members to co-edit the same document simultaneously. This is particularly useful for activities that require students' input to be collected instantly.

A Classroom Example of Using Google Docs

Here is an example of how Google Docs can be used in a class to collect students' inputs in real time. In this class, the 24 year-one university student teachers in NIE were tasked to learn eight pedagogical approaches that can be used to integrate ICT into teaching and learning. It is almost impossible to achieve such task by individual effort. Hence group effort and a proper tool to facilitate the group work were needed. To do this, the student teachers were divided into eight groups of three where each group was given only one approach to discuss and learn. Group members could use Internet search to help gather information and then collectively and collaboratively they had to provide findings in terms of the following: the focus of the approach, the theories behind the approach, the characteristics of the approach, the ways the approach is applied in schools and examples to demonstrate how the approach is applied. The essence of the activity was to learn by way of "Distributed Cognition" where the learning task is distributed among many learners and the findings from each learner are later shared among the learners (Bell and Winn, 2000). The aim is to learn efficiently by maximizing resources. The tool to facilitate the group activity is Google Docs. To use the tool, the tutor had to prepare beforehand a word document table that formed a two-dimensional matrix with the rows containing the pedagogical approaches and the columns containing the findings (see Figure 5). This table was copied to the Google Docs page for all class members to access. Each class member was issued editor rights prior to the activity.

During the lesson, each group of student teachers logged in to the Google Docs and opened the table file. Each group then researched on the approach through discussion and Internet search. They then fill in their findings on the table. The table entry can be done at any time simultaneously. At the end of the activity, the table was filled and he tutor projected the table to the whole class. This provides two prongs of learning: first, the tutor uses the projection to point out possible mistakes and areas of concern; second, the student teachers can continue to make amendment to their findings and view their peers' findings from other groups. Figure 5 shows a screen shot of the Google Docs completed by the class.

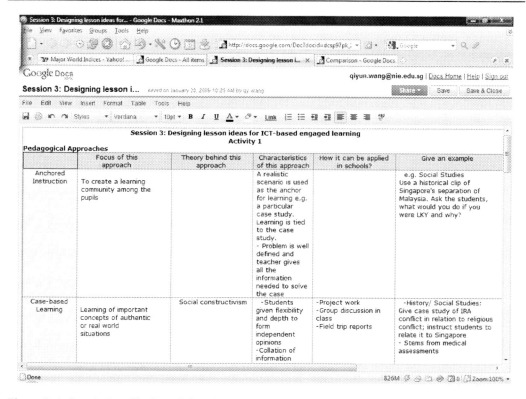

Figure 5. A Google Doc file for collaborative editing.

Besides allowing users to co-edit a Google Docs document, a Google Docs shares almost similar properties as the PbWiki. Therefore, a Google Docs is more suitable for activities that require the students' input immediately whereas a PbWiki is good for activities that require more time for thinking and reflection.

Yahoo Groups

A Yahoo group is a congregation of people of common interests to perform activities in a common cyberspace such as sharing of ideas, photos and files. It also provides functions like group announcement and discussion forum. For example, when one member wants to send a message to all members, he/she needs only to post the message to the specific Yahoo group he/she belongs to; and the message will automatically be forwarded either via emails or message posts to all the members. Recipients of the message can reply by the same means, either email or message post.

Going by the features of what a Yahoo group can afford, a Yahoo group can also be used to support collaborative learning effectively. One main difference between a Yahoo group and the other two co-editing tools, the PbWiki and Google Docs, is that a Yahoo group does not support co-editing of a document but rather it allows only uploading and downloading of files and other documents. In other words, it supports only document repository. Therefore, to fully utilize Yahoo group affordances, the activity must be designed on the basis of sharing and a provision for discussion in order to learn collaboratively.

A Classroom Example of Using a Yahoo Group

Here is an example of how a Yahoo Group was created to facilitate a group of 24 pre-service student teachers in NIE who were doing their service learning project for an autism center in Singapore.

This project lasted for a year. The aim of the project is to have the student teachers learn what service learning is all about through taking part in community works. Such a project requires good planning and hence they need to have many meetings. Student teachers found arranging a common time and place for the meetings were difficult and the meetings were often not productive because members usually complained that they had other commitments and the meetings were usually ended abruptly. To resolve this problem, the student teachers created a Yahoo group in the Internet to allow all members to interact with each other without having to worry about logistic and time schedule issues.

They even invited the group supervisor to join the group as a member so that they could receive just-in-time guidance. Their interactions include posting suggestions, taking part in discussion forums to brainstorm for ideas and using online voting function to make decision. In the whole process, members altogether posted 97 messages, uploaded 12 files, 1 folder, 1 photo album with 10 photos inside, 1 link and initiated 2 sessions of polling. Just in the first two months, they already posted 69 (=71%) messages. These data show that the members indeed used the Yahoo group quite extensively to get their tasks done. Figure 6 shows a screen capture of the messages posted in the Yahoo group.

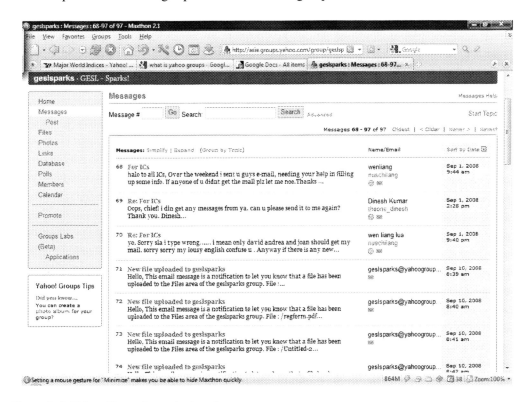

Figure 6. A Yahoo Group for project work.

Facebook

Facebook is a social networking platform through which users can keep in touch with old friends, make new friends, share resources such as photos, videos, news and create group pages to update others on their latest happenings. Compared with the rest of the tools discussed above, Facebook works in an open environment with the intention to reach out to as many people as possible. So an invited friend in a Facebook may bring along his/her own circle of friends which is called "friends of friends". So a Facebook can be easily populated by this mechanism.

Facebook works by first requiring a potential subscriber to apply for a personal account. The account holder is then free to publish any information on his/her own Facebook. He/she then invite others to view his/her Facebook by adding the invitees as "friends". Once becomes friends, the information published by the inviter will be able to be viewed by all the friends as well as the "friends of friends". This is one powerful way that Facebook can reach out to a large body of users.

Facebook can help to establish and maintain immediacy among students and between the students and the teacher. Here immediacy refers to the psychological closeness between two partners. Compared to a face-to-face classroom, interaction in an online learning environment normally lacks the social cues such as gesture, eye contact, and other emotional expressions. The lacking of these cues can potentially pose challenges to forming rapport and building good relationship. Facebook, being a social networking tool, provides an alternative way to increase social cues. For instance, Facebook allows users to upload photos and videos and invite viewers to share their views. The process is equivalent to asking for opinion in a face-to-face meeting. Facebook also affords users to record a video through a webcam and upload the video immediately to a Facebook page. The instant capturing of video images is useful for education when just-in-time information is crucial in the learning process.

One unique feature of Facebook is for individual to self-disclose his/her life style and the state of doing called "profile" in Facebook. Such self-disclosure when administered properly can act as a motivator to "lure" students into Facebook to sustain the interactivity within the Facebook community. This is especially so if the self-disclosure involves the teacher. Research finds that a teacher's self-disclosure can help create positive relationship between the teacher and the students. This positive relationship will in turn help to maintain supportive class climate. Similarly the self-disclosure of students in Facebook may also help students to become close friends or help group members to know each other better which translate into better team bonding and hence better group output.

Other features in a Facebook include using *Wall* to update friends on one's own happenings. Items put on one's wall will not be broadcast outward so it retains the privacy to only close friends. On the other hand, if the same item is to be sent using *Notes*, it will be broadcast to all friends and the item will appear in the friend's Facebook home page – which is the first page that each Facebook will see upon logging in. Because of this provision for choice of recipient, Facebook can be used to address a selected group of students depending on the teacher's choice. Another very useful function in Facebook is *Group*. Group allows the owner to form a special group with a specific purpose such as a focus group or a project group in a school setting. Members in the Group can take part in a discussion forum meant only exclusively for the group members. The Group also provides a function for sending out announcement to inform others about a forthcoming event. Therefore, the Group function is very suitable for close environment interaction.

A Classroom Example of Using a Facebook

Here is an example of how a Facebook group was used to support a group of Master degree students in an elective course on designing e-learning tools for teaching and training. Five students were involved in the course. The students meet once a week and the course lasted for 13 weeks.

A Facebook group MID822_Jan09 was created to facilitate the group collaborative learning. Two discussion forums were created within the group page. See Figure 7.

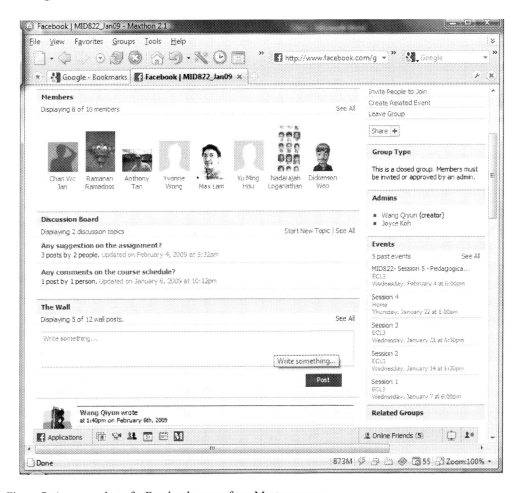

Figure 7. A screen shot of a Facebook group for a Master course.

Participants use the discussion forum to share ideas and obtain feedback on their assignments, course materials and other issues they face during the course of study. The tutor used the *Event* function to make announcement for every session and also the instruction for their class activities. The instruction also includes all necessary learning materials like photos, videos and links. Students used the "Share" function located inside the Event page to make comment with regard to the learning materials and solicit reply from other group members.

As an illustration, the Group was used as a platform to launch elearning activities for the fourth week because the fourth face-to-face session was cancelled due to a public holiday. Therefore, the missed session was replaced by a week of elearning activities. The elearning activities basically consisted of reading tasks and group learning. Students first read a few articles given by the tutor and then posted a summary together with a reflection on what they had read on the Wall inside the Event page. They then wrote comments, critiqued their friends' work and replied to queries if need be. The purpose of the activity was to get students learn with the Zone of Proximal Development (ZPD) (Slavin, 1997) so as to maximize their learning potential with the help of peers.

CONCLUSION

The ability to collaborate in a learning process has become an essential competency for people in the new information age and knowledge society. People must learn and know how to work with others so that they can together solve complicated and authentic problems productively. However, simply by putting people together cannot guarantee desirable output and work harmony. The environment must be conducive for group interaction and the communication means must be convenient. Certain internet tools and strategies are useful in this aspect and are known to have facilitated collaborative learning well.

The web 2.0 tools introduced in this paper, that is, Weblog, Wiki, Google Docs, Yahoo Groups, and Facebook are derived from emerging technologies that have attracted great attention in recent years. The strengths of these tools are that they support two-way communication, facilitate interaction and most importantly, they promote authorship autonomy and accountability. These are great ingredients for making collaborative learning a success (Wang, 2009). In all, although web 2.0 tools offer great promise to support collaborative learning, the true success of their usage lie on the sound pedagogies that were behind the design of the activities, hence, the tool alone cannot make wonders but the designers and the users certainly can make a difference.

REFERENCES

[1]   Barron, B. (2000). Achieving coordination in collaborative problem-solving groups. *The Journal of the Learning Sciences, 9*(4), 403-436.
[2]   Bell, P., and Winn, W. (2000). Distributed cognitions, by nature and by design. In D.H. Jonassen, and S.M. Land (Eds), *Theoretical foundations of learning environments* (pp. 123-145). NJ: Lawrence Erlbaum Associates.
[3]   Haythornthwaite, C. (2006). Facilitating collaboration in online learning. *JALN*, 10(1), 7-24.
[4]   Lipponen, L., Hakkarainen, K., and Paavola, S. (2004). Practices and orientations of CSCL. In J.W. Strijbos, P.A. Kirschner, and R.L. Martens (Eds.), *What we know about CSCL* (pp. 31-50). Norwell, MA: Kluwer Academic Publishers.
[5]   Neo, M. (2003). Developing a collaborative learning environment using a web-based design. Journal of Computer Assisted Learning, 19(4), 462-473.

[6]   Slavin, R. (1997). *Educational psychology: theory and practice*. MA: Allyn and Bacon.

[7]   Wang, Q.Y. (2009). Editorial. *International Journal of Continuing Engineering Education and Life-long Learning, 19*(2/3), 107-111.

[8]   Wang, Q. Y., and Woo, H. L (2008a). The affordances of weblogs and discussion forums for learning: A comparative analysis. *Educational Technology, 48*(5), 34-38.

[9]   Wang, Q.Y., and Woo, H.L (2008b). Affordances and innovative uses of weblogs for teaching and learning. In Kobayashi, R (Ed.), *New Educational Technology* (pp. 183-199). NY: Nova Science Publishers.

INDEX

B

C

D

E

F

G

J

N

O

P

Q

R